Co...

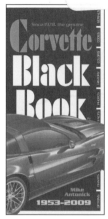

Send _____ copies @ $17.99 each $ _____

Ohio residents add 1.25 sales tax _____

Postage/hard shipping container **3.00**

Check or money order enclosed $ _____

Mail order to:
Michael Bruce Associates, Inc
Post Office Box 1966
Gambier, Ohio 43022

Name _____

Street _____

City _____ State _____ Zip _____

BLACK BOOK ORDER FORM

Corvette Black Book 1953-2009

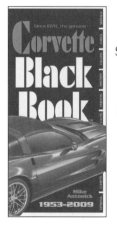

Send _____ copies @ $17.99 each $ _____

Ohio residents add 1.25 sales tax _____

Postage/hard shipping container **3.00**

Check or money order enclosed $ _____

Mail order to:
Michael Bruce Associates, Inc
Post Office Box 1966
Gambier, Ohio 43022

Name _____

Street _____

City _____ State _____ Zip _____

BLACK BOOK ORDER FORM

BLACK BOOK ORDER FORM

2

The Corvette Black Book
1953-2009

Co-published by

Michael Bruce Associates,Inc.
Post Office Box 1966
Gambier, Ohio 43022
www.corvetteblackbook.com

MBI Publishing Company
400 First Avenue North
Suite 300
Minneapolis, MN 55401 USA

Contents

This edition published in 2009 by Michael Bruce Associates, Inc., and Motorbooks an imprint of MBI Publishing Company, 400 First Avenue North, Suite 300, Minneapolis, MN 55401 USA.

Copyright © 2009 by Michael Bruce Associates, Inc..

All rights reserved. With the exception of quoting brief passages for the purposes of review, no part of this publication may be reproduced without prior written permission from the Publisher.

The information in this book is true and complete to the best of our knowledge. All recommendations are made without any guarantee on the part of the author or Publisher, who also disclaim any liability incurred in connection with the use of this data or specific details.

This publication has not been prepared, approved, or licensed by General Motors Corporation.

We recognize, further, that some words, model names, and designations mentioned herein are the property of the trademark holder. We use them for identification purposes only. This is not an official publication.

Motorbooks titles are also available at discounts in bulk quantity for industrial or sales-promotion use. For details write to Special Sales Manager at MBI Publishing, 400 First Avenue North, Suite 300, Minneapolis, MN 55401 USA.

To find out more about our books, visit us online at www.motorbooks.com.

We acknowledge with appreciation the following who contributed their expertise to this and previous *Corvette Black Books*: Noland Adams, Dan Aldridge, John Amgwert, Pat Antonick, Bob Applegate, Pat Baker, J. Scott Barnes, Jane Barthelme, Sanford Block, Michele Boling, Kent Brooks, Barry Brown, Dale Brown, David Burroughs, Jerry Burton, Harlan Charles, Ladd B. Chase, Alan Colvin, John Coyle, Steve Dangremond, M. F. Dobbins, Zora Duntov, Bob Eckles, Sam Folz, Floyd Foust, Fred Gallasch, Steve Guckenberg, Dick Guldstrand, Joe Haase, Hib Halverson, John Hibbert, Dave Hill, Bill Hudak, Mike Hunt, John Hyland, Rick Johnson, Patrick Johnston, Tadge Juecter, Alan Kaplan, Othar Kennedy, Gordon Killebrew, Paul Kitchen, Gary Konner, Jim Krughoff, Michael LaEnvi, Rob Lindsay, Gary Lisk, Bill Locke, Bob Lojewski, Kevin Maley, Michael Marks, Bob McDorman, Dave McLellan, Chip Miller, Bill Mock, Bill Munzer, Bill Nichols, John Osborn, Brian Pearce, John Poloney, Wes Raynal, Bill Rhodes, Ray Rocque, Andy Roderick, Chuck Ryan Jr, Ken Schurr, Donna Sims, Stan Slade, Jeffrey Smith, Mark & Dixie Smith, Dennis Tracy, Lou Vitalle, Jerry Wadsworth, Jerry Weichers, Don Williams, Jon Yaeger, Mike Yager, and Ray Zisa. Thanks also to Chevrolet, Callaway Engineering, and to Mercury-Marine.

Notice 1: *The Corvette Black Book* and its publisher, Michael Bruce Associates, Inc. have no relationship or connection whatever with Hearst Business Media Corporation, its parent or affiliated corporations, or the *Black Book* published by National Auto Research Division of Hearst Business Media Corporation.

Notice 2: Michael Bruce Associates, Inc. and the *Corvette Black Book* are not associated with or sponsored by GM or its Chevrolet Motor Division.

Cover: Front cover photo courtesy of Chevrolet; back cover photo by Gary Lisk; cover design by Mike Antonick

Sponsor: The *2009 Corvette Black Book* is sponsored by Bob McDorman Chevrolet, Canal Winchester (Columbus), Ohio. 1-888-207-1865

Printed and bound in the United States by Cushing-Malloy, Inc., October 2008.

ISBN-13: 978-0-7603-3602-1

Glossary of Terms

ABS: Anti-lock Braking System. Sensors monitor wheel rotation and prevent wheel lockup during braking by modulating hydraulic pressure. Corvette's ABS was standard equipment starting with 1986 models.

ASR: Acceleration Slip Regulation. Engine spark retard, throttle close down, and brake intervention limited wheel spin during acceleration. ASR, or traction control, was standard for Corvettes starting with 1992 models.

A-Arm: A lateral suspension link resembling the letter A.

A-Pillar: The post section between the windshield and door glass.

Active Handling: A suspension system with actuators at each wheel controlled by an electronic control unit. Corvette's active suspension was optional from 1998-2000, then standard starting in 2001.

Airbag: An inflatable bag deployed in front or side impact intended to cushion and restrain a vehicle's occupants. First installation of an airbag in Corvette came in 1990.

Analog: In instruments, the classic design which typically used a pointer moving along an arc to display information. Corvette's main instruments were analog through 1982, partially analog from 1990 through 1996, then full analog again starting in 1997.

Anniversary Edition: Special paint and/or trim to mark the anniversary model years of 1978 (25th), 1988 (35th), 1993 (40th), and 2003 (50th).

Aspect Ratio: The ratio of a tire's cross-sectional height to its width.

Backlite: An automobile's rear window.

B-Pillar: In Corvettes, the area between the door glass and rear window.

Benchmark: Bloomington award for gold certified and survivor combo.

Big Block: Large displacement V8 engines of either 396, 427 or 454 cubic inches, optional in Corvettes from 1965 through 1974.

Big Tank: Optional thirty-six gallon fuel tank for 1963 through 1967, or twenty-four gallon fuel tank for 1959 through 1962 Corvettes.

Billy Bob: Chevy engineers' unofficial code name for the 1999 hardtop model, which was initially planned to be a low-content, price leader.

Black Hills: Popular Corvette street show in Spearfish, South Dakota.

Blueprint: Precise finishing of an engine's components, often by hand, to optimal size within factory specifications.

Blue Devil: Unofficial code name for the 2009 ZR1 model.

Blue Flame: Six cylinder engine for 1953, 1954 and a few 1955 models.

Bloomington: A major annual Corvette show, known for its certification judging system. Held in Bloomington, Illinois from 1973 through 1992; in Springfield, Illinois from 1993 through 1996; back to Bloomington, Illinois from 1997-2001, then to Pheasant Run Resort, St. Charles, Illinois in 2002.

Body Off: Extensive restoration in which a Corvette body is completely removed from its frame.

Bolt On: Optional 1967 cast aluminum road wheel.

Bose: Integrated radio and speaker system introduced optionally in 1984.

Bowling Green: Location of the Corvette assembly plant since 1981.

Broadcast Code: Designation for the two or three character alpha code stamped, etched or otherwise affixed to components which identifies usage. Enthusiasts call the engine broadcast code the *Suffix*.

Bucket: Headlight container for 1953, 1954 and 1955 models, 1963 to 1996 models, or the bucket seat for any model year.

Bullet: Individual air inlet for 1953 and early 1954 Corvettes.

Buzzer: Tachometer rpm limit warning device used briefly in 1963; or the similar speedometer warning device optional in 1967-1969.

CAFE: Corporate Average Fuel Economy.

Callaway Twin Turbo: Engine conversion by Callaway Engineering, Old Lyme, Connecticut. Available from select Chevy dealers from 1987-1991.

Carlisle: Site of several auto events including a major annual Corvette

Glossary...continued

show at the Carlisle, Pennsylvania fairgrounds in late August.

Casting Number: Usually refers to the GM part number cast into engine blocks, but can refer to cast numbers on other components.

Challenge Cars: Corvettes factory-built for the all-Corvette race series sanctioned by the Sports Car Club of America (SCCA) in 1988 and 1989.

CHMSL: Center high-mounted stop lamp.

China Wall: The surface between the intake manifold and cylinder block.

Classic: 1953 through 1962 model Corvettes.

Coefficient of Drag: A calculation of airflow over a vehicle's surface, usually measured in a wind tunnel.

Collector Edition: Specially equipped and trimmed 1982 or 1996 models.

Commemorative Edition: Special LeMans Blue paint and trim 2004 models to commemorate Corvette's LeMans racing success.

Convertible: Soft top Corvettes built between 1953-1975, and starting again in 1986. Some enthusiasts prefer the term *roadster* for 1953-1955 models.

Cove: Concave section of front fender and door area introduced with 1956 models which could be painted a contrasting color (optional). Part of the area became convex in 1958. A contrasting color option was not available in 1962, the cove's last year.

Cowl: The surface of a car between the windshield and hood, bounded on the sides by the fenders.

COPO: Central Office Production Order.

Corvette Black Book: Pocket "bible" of Corvette facts.

Coupe: 1963-1967 fixed-top Corvettes; also T-top 1968-1982 models, and removable-roof panel (targa) 1984 and newer models.

Curb Weight: The road-going weight of a vehicle including fluids, but excluding occupants or cargo.

Cypress: Major January NCRS Corvette show in Cypress Gardens, Florida; moved to Disney World in 1999, then to Kissimmee, Florida in 2000.

C1,C2,C3,C4,C5,C6: Abbreviations for Corvette "generations" which came into popular use in the 1990s: C1=1953-1962, C2=1963-1967, C3=1968-1982, C4=1984-1996, C5=1997-2004, C6=2005+.

Digital: In instruments, the presentation of data in numeric form. Corvette's main instruments were digital from 1984-1989, partially digital from 1990-1996.

Doghouse: Fuel injection plenum chamber, 1957-1965 models.

DOHC: Dual overhead camshafts.

DOT: Department of Transportation.

Drive-by-Wire: A system of electronic accelerator activation to replace mechanical linkage. Corvette adapted drive-by-wire starting in 1997.

DRL: Daytime Running Lights.

Duntov: Zora Arkus-Duntov, legendary chief engineer, called "father" of the Corvette; also, National Corvette Restorers Society (NCRS) award combining static judging with mechanical excellence.

Drum Brake: All Corvettes built before 1965 have them, but the term sometimes refers to the few 1965 models built without disc brakes.

Elephant Ears: Rubberized canvas front brake air scoops for some 1957 through 1964 Corvettes with heavy-duty racing brakes.

EPA: Environmental Protection Agency.

Fiberglass: A material combining thermoset plastic with glass fibers.

FOA: Factory optional accessory.

Flint: Michigan site of first Corvette assembly facility. Also, site of GM engine plant which built most Corvette small block engines through 1996.

Four+Three: Doug Nash four-speed manual transmission with three overdrives available in Corvettes from 1984 through 1988.

Fuelie: 1957 through 1965 fuel-injected Corvette.

EMT: Extended Mobility Tire (run flat).

Glossary...continued

Glider: Corvette assembled at St. Louis with no engine or transmission.

Gold Certified: Top Bloomington award for factory originality.

Gold Line: Optional gold stripe tire for 1965-1966 models.

Gold Spinner: Top award at Vettefest show in Illinois.

Grand Sport: Five 1963 factory-built racers; also 1996 option package.

Gymkhana: Optional 1974 through 1982 suspension package.

Hardtop: Removable auxiliary hardtop available for 1956-1975 and 1989-1996 convertibles; also new fixed-roof model introduced in 1999.

Head-up Display: Projection of instrument readout onto a vehicle's windshield in the direct vision path of the driver. Corvette first offered this feature in its 1999 model.

HVAC: Heating-ventilation-air conditioning system.

King of the Hill: Media terminology for the 1990 Corvette ZR-1.

Knock Off: Cast aluminum wheel optional for 1963 through 1966 models. Evidence suggests 1963 was over-the-counter only.

Knoxville: Annual Corvette show in Knoxville, Tennessee.

Lateral Support: In seating, the restraint of side to side movement.

Lumbar Support: In seating, support for the torso's lower lumbar region.

LPO: Limited Production Option.

LT1: High performance engine option for 1970, 1971 and 1972. Also, redesigned base engine from 1992 through 1996.

Magnesium: Weighing less than aluminum, it's used where low weight is required. A magnesium road wheel was optional for 1997-2004 Corvettes. The 2006 Z06 intruduced a magnesium-supported fixed roof panel.

Mako Shark: A Corvette showcar.

Matching Numbers: Vin-derivative stamped into engine matches VIN; also can refer to any stamped or cast codes indicating original parts.

Mid Year: 1963 through 1967 Corvette.

Mouse Motor: Slang term for "small block" Chevy engines.

Multiport: Technically accurate description for the fuel injection of 1985 and newer models, but especially for 1992 and newer.

NACA Duct: A design for a low restriction air inlet developed by the National Advisory Committee for Aeronautics, it is often seen on race cars, but was also used on the hoods of 1987 Callaway Twin Turbo Corvettes and on some one-off Corvette showcars.

NCCC: National Council of Corvette Clubs.

NCM: National Corvette Museum, located in Bowling Green, Kentucky.

NCRS: National Corvette Restorers Society.

NOS: New old stock (brand new old parts).

NHTSA: National Highway Traffic Safety Administration.

OEM: Original Equipment Manufacturer.

Over The Counter: Parts sold at Chevrolet dealers' parts departments.

Pace Car: Limited edition replicas of Indianapolis 500 pace cars offered in 1978, 1986 (all convertibles), 1995, 1998 and 2007. Corvette also paced the race in 2002, 2004, 2005 and 2006.

Pilot: Pre-production vehicle built on factory assembly line.

Plexiglas: Trade name for a transparent plastic material often used to replace glass in automotive applications.

Prefix: Alpha stamping into engine identifying engine build plant.

Rat Motor: Slang term for "big block" Chevy engines.

Red Line: Optional 1967-1969 red stripe tire. Also, maximum tach rpm.

Registry: An organization of owners of a specific year, model or series of Corvettes for the purpose of registering and sharing information about their vehicles.

Replacement: Part furnished by Chevrolet which fits and functions, but doesn't necessarily duplicate the original.

Glossary...continued

Roadster: Enthusiast (not Chevrolet) term for 1953 through 1955 Corvette.

Rock Crusher: Heavy-duty, close ratio 4-speed manual transmission (RPO M22) available between 1966 and 1971.

Romulus: Engine plant source for 1997-1999, and most 2000 models.

RPO: Regular Production Option.

SAE: Society of Automotive Engineers.

St. Catharines, Ontario, Canada: Engine plant source for base-engine Corvettes starting in 2000.

St. Louis: Site of Corvette assembly factory from 1954 through 1981.

Selective Ride: Adjustable shock absorber system first available in 1989.

Serpentine: A flat, grooved belt that drives several components. A single serpentine belt typically replaces several individual belts.

Shark: A Corvette showcar; also Corvette production models from 1968 through 1982.

SIR: Supplemental Inflatable Restraint (airbag) introduced in 1990.

Side Curtain: A detachable window consisting of clear plastic. In Corvette's case, 1953-1955 models had this type of side window.

Side Pipes: Optional 1965 through 1967, and 1969 side-mount exhausts.

Silver Anniversary: Two-tone silver 1978 paint option.

Small Block: V-8 engines of 265, 283, 305 (1980 California only), 327 and 350 cubic-inch displacement.

Solid Axle: 1953 through 1962 Corvette.

Split Window: 1963 Corvette coupe.

Sticker: Price posted on government-required window sticker; retail.

Stingray: 1969 through 1976 Corvette.

Sting Ray: 1963 through 1967 Corvette.

Suffix: Two or three character alpha code stamped or etched into an engine block which identifies usage. Also called the Broadcast Code.

Survivor: Bloomington award for unrestored, mostly original Corvette.

Sway Bar: A transverse bar linking suspension components. May be solid or tubular. Also called an anti-roll bar, or anti-sway bar.

Synthetic Oil: A lubricant which is chemically engineered to have superior properties to oil refined directly from natural crude. Mobil 1 brand synthetic oil has been standard factory-fill for Corvettes since 1992.

Targa: A roof design for coupes which has a single removable panel. Corvette coupes have been targa designs since the 1984 model.

T-top: A coupe design pioneered by Corvette with two removable roof panels and a fixed center structural member. It was used for 1968 through 1982 coupe models.

Teak: Teakwood steering wheel optional in 1965 and 1966.

Tonawanda: GM engine plant near Buffalo, New York.

Top Flight: NCRS judging award for factory originality.

Tumblehome: Viewed from front or rear, the inward curvature of a vehicle's side from the beltline upward.

Vettefest: Major annual Corvette show in Illinois.

Tuned Port: 1985 through 1991 fuel injection with "tuned" runners.

VIN: Vehicle Identification Number.

Wheelbase: The distance between wheel centers when view from the side.

Yaw: To turn by angular motion about the vertical axis.

Zero Offset Steering: Steering system geometry with zero scrub radius. Corvette adopted zero offset steering with its 1988 model.

Z06: High-performance option for 1963 models; also high-performance hardtop-derived 2001-2004 C5 model; also high-performance C6 coupe-derived model introduced in 2006.

ZR1: 1970-72 engine option; 2009+ supercharged Corvette model.

ZR-1: 1990-1995 Corvette model with 32-valve, overhead cam engine.

Historic Dates

December 25, 1909: Zora Arkus is born in Brussels. He is the son of Russian parents, Rachel and Jacques Arkus, who were students in Belgium. He later modified his surname to Arkus-Duntov.

November 3, 1911: Chevrolet Motor Division of General Motors is formed.

January 1, 1928: Art and Colour, the industry's first dedicated styling department, is created by GM. Californian Harley Earl is named its head.

June 2, 1952: Harley Earl gives Chevrolet Chief Engineer Ed Cole a sneak preview of an Art and Color mockup of a secret two-seat sportscar, codenamed Opel. Cole, in his Chevrolet position for only a month and determined to inject excitement into the staid division, enthusiastically seeks permission to produce the vehicle as a Chevrolet product.

January 17, 1953: The new sportscar meets the public for the first time at the General Motors Motorama held at the Waldorf-Astoria hotel in Manhattan. From 1,500 names submitted, Cole has selected *Corvette*.

May 1, 1953: Zora Arkus-Duntov is hired by Chevrolet. After rejections from GM, Ford and Chrysler, Duntov redoubled his efforts to join GM after seeing the Motorama Corvette. However, he is hired as an assistant staff engineer for engine development with no Corvette responsibilities.

June 30, 1953: The first production Corvette rolls off the assembly line at a temporary facility on Van Slyke Avenue in Flint, Michigan.

September 28, 1953: Banned from working on the Corvette, Zora Arkus-Duntov nonetheless presents his views on sportscars in an address to the Society of Automotive Engineers (SAE) in Lansing, Michigan.

December 16, 1953: Arkus-Duntov drafts an internal Chevrolet memo titled, "Thoughts on Youth, Hot-Rodders and Chevrolet." In it, he lays out the high-performance image he believes Chevrolet and Corvette should develop, and forever links himself to the effort.

December 28, 1953: Production begins on the 1954 Corvette at a freshly renovated facility in St. Louis, Missouri. Demand is estimated at 10,000 units annually, but actual sales fall far short.

December 20, 1955: Using a V-8 powered 1954 test mule, Zora Arkus-Duntov exceeds 156 mph at GM's Mesa Provings Grounds.

January 1956: Zora Arkus-Duntov starts to burnish the Corvette's performance image with a NASCAR sanctioned and timed two-way average speed of 150.583 mph for the flying mile on the sand at Daytona Beach, Florida. He drives the redesigned 1956 model, introduced to the public the same month at the GM Motorama in New York City. One key to Duntov's success is a new camshaft of his own design.

February 18, 1956: John Fitch arrives in Sebring, Florida to begin preparations for the entry of four Corvettes into the 12-hour Sebring race just five weeks away. Fitch, possibly the finest American roadracer of his era, was hired by Ed Cole who was energized by Duntov's success (and the resulting publicity) following the Daytona Beach runs.

March 24, 1956: "Less than we hoped for, but more than we deserved," was how John Fitch summarized Corvette's assault on Sebring as two of the four entries finished the race in positions of ninth and fifteenth overall.

December 1, 1958: Harley Earl retires as head of GM Design Staff. His successor is William L. Mitchell, the son of a Pennsylvania Buick dealer.

March 23, 1957: Corvette returns to Sebring. Duntov has created a purpose racer, the Corvette SS, with spectacular styling. John Fitch and Piero Taruffi retire it after 23 laps with a broken suspension bushing.

June 1957: The Automobile Manufacturers Association suggests its members stop active participation in automobile racing. GM agrees.

April 18, 1959: The Sting Ray racer appears at Maryland's Marlboro Raceway. It is privately entered by William L. Mitchell and driven by the "flying dentist," Dr. Dick Thompson. Using the chassis from a left over SS Corvette test mule, Mitchell directed the design of a spectacular new body which will lend its design theme to production Corvettes in 1963.

April 10, 1960: John Fitch and Bob Grossman bring their #3 Corvette home in eighth place overall at the 24-hours of LeMans.

October 7, 1960: The *Route 66* television show debuts, starring Martin Milner, George Maharis, and...Corvette.

February 18, 1961: The Sting Ray Racer debuts as a show car at the Chicago Auto Show. It has retired from racing, and has been refinished to perfection.

October 12, 1962: At Riverside, Doug Hooper wins the production 1963 Sting Ray's first race, but only because mechanical problems keep Carroll Shelby's Cobras, also in their first race, from finishing.

April 1, 1967: The L88 big block sees its first race action at the 12-Hour Sebring event, winning its GT Class and placing 10th overall.

November 19, 1969: The 250,000th Corvette is built, a Riverside Gold convertible.

October 4, 1973: The Four-Rotor Corvette show car, later renamed the Aero Vette, debuts at the Paris Auto Show.

December 31, 1974: Zora Arkus-Duntov retires.

January 2, 1975: David McLellan is appointed Corvette chief engineer.

March 15, 1977: St. Louis builds the 500,000th Corvette, a white T-top coupe with red interior.

May 7, 1977: Ed Cole, retired as GM President for two years, dies at the controls of his private plane near Mendon, Michigan.

May 26, 1978: A Corvette, driven by Jim Rathman, paces the Indianapolis 500 auto race. Replicas set off a fierce, but short lived, buying frenzy.

April 22, 1980: Dave McLellan makes a formal presentation for approval to design and build an all-new generation of Corvettes, tentatively scheduled for 1983. GM's Product Policy Group approves.

June 1, 1981: A new Corvette assembly plant in Bowling Green, Kentucky, completes its first build.

July 31, 1981: The St. Louis Corvette assembly plant builds its last Corvette, VIN #1G1AY8764BS431611, a beige coupe.

December, 1982: The fourth-generation Corvette is introduced to the media at Riverside International Raceway. It will be a 1984 model.

October 26, 1983: Bowling Green builds the 750,000th Corvette, a white coupe.

October 1985: Corvette reintroduces a convertible model to the motoring press at Yosemite National Park in California.

May 25, 1986: A Corvette, driven by aviator Chuck Yeager, paces the Indy 500 auto race. All 1986 model convertibles are designated as Pace Car replicas.

March 7, 1989: The 1989 Corvette ZR-1 debuts at the Geneva Auto Show.

April 19, 1989: Chevrolet notifies its dealers that, due to insufficient engine availability, the ZR-1 will be delayed and released as a 1990 model.

March 2, 1990: A ZR-1 with a stock LT5 engine sets new FIA 24-hour speed and endurance records at Firestone's Fort Stockton, Texas, test oval.

June 5, 1992: Ground is broken for the National Corvette Museum in Bowling Green, Kentucky.

July 2, 1992: Bowling Green builds the 1,000,000th Corvette, a white convertible. Chevrolet throws a birthday bash in Detroit five weeks later.

August 31, 1992: Dave McLellan accepts early retirement.

November 18, 1992: Dave Hill becomes the Corvette's third chief engineer.

November 16, 1993: Mercury Marine completes LT5 engine construction, though ZR-1 vehicle assembly will continue through the 1995 model year.

September 2, 1994: The National Corvette Museum opens.

May 28, 1995: A 1995 Corvette convertible, driven by Chevrolet General Manager Jim Perkins, paces the Indianapolis 500 auto race.

April 21, 1996: Zora Arkus-Duntov dies in Grosse Pointe, Michigan.

May 24, 1998: A 1998 Corvette convertible, driven by Parnelli Jones, paces the Indianapolis 500 auto race.

May 26, 2002: A 2003 Corvette 50th Anniversary coupe, driven by actor Jim Caviezel, paces the Indianapolis 500 auto race.

May 30, 2004: A 2004 Corvette convertible, driven by actor Morgan Freeman, paces the Indianapolis 500 auto race.

May 29, 2005: A 2005 Corvette coupe, driven by former Secretary of State Colin Powell, paces the Indianapolis 500 auto race.

December 31, 2005: Dave Hill, Corvette's third chief engineer, retires.

January 2, 2006: Tom Wallace, Corvette's fourth chief engineer, assumes duties.

May 28, 2006: A 2006 Corvette Z06, driven by seven-time Tour de France winner Lance Armstrong, paces the Indianapolis 500 auto race.

May 27, 2007: A 2007 Corvette Atomic Orange convertible, driven by Patrick Dempsey, paces the Indy 500 auto race. Five-hundred replicas are sold.

May 25, 2008: A 2008 Corvette Z06, modified to use E85 ethanol and driven by Emerson Fittipaldi, paces the Indianapolis 500. 500 replicas of a second pace car design, black and silver to commemorate the 1978 pace car, are sold.

Instructions

The *Corvette Black Book* is intended to help you understand and enjoy Corvettes. It does so by presenting useful and interesting data in a readily accessible format. Perhaps more so than any other auto enthusiast group, Corvette owners are conscious of details. You've seen the term "numbers match" associated with Corvettes. This book will help you understand what those numbers are and what they mean.

But numbers are only part of the story, part of an individual Corvette's history. Each Corvette model year has scores of unique features. Within a model year, the range of colors and options available, except for the first couple of years of Corvette production, permitted thousands of combinations. For most Corvette model years, it was theoretically possible that no two Corvettes were built with exactly the same combination of options, colors and equipment. That wasn't the case, of course, but it was possible. Understanding how to determine what is rare and what isn't is another part of understanding an individual Corvette's mystique. Again, the *Corvette Black Book* can help.

Before explaining how to interpret the data presented here, some words of caution. This book was first published in 1978. It has been updated and refined many times since, often several times within a year. It's done by starting with Chevrolet records, documents and personnel, then soliciting the critiques of expert Corvette enthusiasts around the country, including dealers, authors, concours judges, owners, restorers and others. Unsolicited mail from enthusiasts, always welcome, sometimes exposes the possibility of an error or an omission. New data or clarifications are always surfacing. In short, it is a never-ending process. This *Corvette Black Book* you are holding is the most complete and accurate yet published. But next year's will be even better.

Yet this book, or any published material, must not be relied upon as a final, definitive guide, especially for determining a Corvette's authenticity or desirability. A high percentage of data presented here is accurate for a high percentage of Corvettes produced, but nothing is absolute in the world of automobiles, especially Corvettes. Deviations are possible in almost every category. In the end, you must balance published data with good judgement, logic and common sense.

Why these cautions? Because of the nature of auto production, a totally accurate listing of automotive facts, particularly relating to part numbers, is impossible. General Motors does not document every production change made. The replacement parts aspect of any auto company is geared to function, not originality. Parts are interchanged during production when shortages occur. New parts inventories are not always phased in so that all "old" parts are used before the new. Frankly, General Motors and other auto companies have had more important things to worry about than documenting their production practices for the benefit of future enthusiasts, restorers, or historians. It wouldn't be nearly as much fun for all of us now if they had.

Former employes of the St. Louis Corvette plant tell of times when the assembly line was about to run out of parts, door trim screws or bumper bolts perhaps, and someone was dispatched to the closest hardware store to find an acceptable substitute so the line could keep rolling. In your analysis of a Corvette, by all means be very critical and conscious of details. There is no question that accuracy and originality are important factors in a Corvette's desirability. At the same time, keep an open mind. Exceptions to practically every rule are possible.

Use this book as an aid in determining originality and correctness. If you find a discrepancy, don't assume something is wrong, only that it may be. Keep looking. What at first appears incorrect may in fact be legitimate, one of only a handful, and thus very desirable. Or it may be the first thread in the unraveling of a completely bogus Corvette. Bogus as in the Corvette that's been resurrected from the dead, with the majority of its components salvaged from other carcasses. There's nothing wrong with such a rebirth, except when it is being represented as an untouched, low mileage original, or anything else it is not.

In some ways, the emphasis on numbers has run amok. There has been so much emphasis on original Corvette parts and components that serious counterfeiting has resulted. As an owner, understand that what you do with your own Corvette is largely your own business. There are laws regarding safety and emissions, of course, but an owner has every right to change colors, interior parts, suspension components and the like. It is the representation of an automobile as something it is not that can constitute fraud. Misrepresentation usually doesn't include the bit of embellishment engaged in by most sellers. It is something like converting a base-engine 1967 Corvette convertible into a 435-horsepower, numbers-matching, low mileage "original" with phony documentation to "prove" it. Be careful in your representations, and in interpreting those of others.

Knowledge is power and the *Corvette Black Book* can put powerful knowledge in your back pocket. The following pages contain graphs of Corvette price and volume trends, and a brief written overview (chronology) of the Corvette's history from 1953 to present. The largest and most useful section is the data section starting on page 26 in which two pages are devoted to each Corvette model year from 1953 to present. A great deal of data is condensed here, and it is important to understand how this information is presented and how to interpret it.

Numbers

The most familiar and widely used number appearing on an automobile is its *vehicle number*, or vehicle identification number (VIN). It may also be called the serial number, or body-chassis number. Each vehicle number is unique. Think of it as an individual Corvette's fingerprint, something shared with no other. It is the vehicle number that appears on current automobile titles and registrations, though in the past some states have used other numbers such as engine numbers.

The vehicle number is assigned prior to vehicle assembly. Each car receives its own number in sequence. The format of the vehicle number for Corvettes has changed over the years, but it always at least indicates the model year and when the Corvette was built relative to others during the same model year. Scrutinizing an individual Corvette's place in the production sequence is simplified by the fact that, except for a two month period in 1981 when Corvettes were built simultaneously at St. Louis, Missouri, and at Bowling Green, Kentucky, all Corvettes during any given model year have been built at one facility.

The vehicle number was stamped or etched into a plate attached to the body of each Corvette. Location varied. From 1953 through early 1960, the plate was attached to the driver-side door post. Most 1960 and all 1961 and 1962 Corvettes had the plate attached to the steering column in the engine compartment. The 1963 through 1967 models had the plate attached to the instrument panel support brace, visible below the glove box. For the benefit of police, federal law required that the vehicle number be visible from outside the vehicle starting in 1968. So, for 1968 and newer Corvettes, the plate was attached to the top surface of the instrument panel, or to the A-pillar (windshield post), both locations visible through the windshield from outside the vehicle. To assist in theft recovery and identification, the vehicle number was also stamped into a Corvette's frame in several locations.

Except for 1955, the vehicle number did not reveal which engine a Corvette had until 1972 when the numbering format was changed to include an engine code. Starting in 1963, all vehicle numbers carried a code differentiating coupe from convertible. From 1999-2004, the hardtop and Z06 hardtop model also carried a separate code.

The *Corvette Black Book* lists the vehicle numbering sequence for all model years from first vehicle produced to last. In most cases, this was a single sequence for Corvettes produced during a model year. There have been exceptions. The 1978 pace car replicas had a separate sequence. In 1981, two plants produced Corvettes simultaneously. 1986 convertibles had separate sequences, as did 1990-1995 ZR-1s. Also, both 1996 Grand Sports 2009 ZR1s had separate sequences.

In earlier years when a single sequence was used for a model year, the last five digits of the last vehicle's number will equal that year's total production. In more recent years, the numbers are not equal due to pilot build or other factors. In 1973, 4,000 vehicle identification numbers were somehow skipped. So total production for 1973 was 4,000 units less than the last five digits of the final vehicle. Also, from 1953 through 1956, the first vehicles produced had vehicle numbers ending with 01001. For these years, final production was 1,000 units less than the last five digits of the last Corvette produced.

It is important to know that Corvette "plants" like Flint, St. Louis and Bowling Green refer to Corvette vehicle assembly (Flint was also the site of an engine plant). Corvette engines have never been built in the same plant as vehicle assembly. Engines were built in dedicated plants which usually built engines for other GM vehicles.

Six-cylinder 1954-1955 engines and the 305 cubic-inch engine used in 1980 California Corvettes were built by Flint Motor in downtown Flint. Big blocks (396ci, 427ci, 454ci) and the six-cylinder 1953 engine were built at GM's Tonawanda, New York, engine plant. The manufacture of the LT5 engine, used in ZR-1s from 1990 through 1995, was subcontracted to Mercury Marine in Stillwater, Oklahoma. All other Corvette small block V8s through 1996 were built at GM's V-8 engine plant in Flint, Michigan. Starting in 1997, Corvette engines were built at GM's plant in Romulus, Michigan. During the 2000 model year, Corvette engines were built at Romulus, and at St. Catharines, Ontario, Canada. All production shifted to St. Catharines for 2001 and later, except for the 2006 and later Z06's LS7 engine, and the 2009 and later ZR1's LS9. The LS7 and LS9 were assembled at GM's Performance Build Center in Wixom, Michigan.

Through 1996, when a Corvette engine was built at an engine plant, an important number was stamped into it. For 1953-1955 six-cylinders, the number was stamped on a machined pad just rear of the ignition distributor opening. For others through 1996, it was on a machined pad just forward of the cylinder head on the passenger side. For 1953, the *prefix* "LAY" preceded a six digit serial number which didn't match the vehicle number. For 1954-1956, a seven digit serial number, not matching the VIN, was followed by an "F" indicating Flint engine build, and two numbers indicating year; for example, F56 for Flint 1956.

Starting in 1957, stamped engine numbers began with an alpha character indicating build plant ("F" for Flint through 1966, "V" for Flint after 1966, and "T" for Tonawanda). The next three or four digits designated date of engine build. The last two (later three) characters were called the *suffix* number (engine plants call it *broadcast code*), even though these digits were alpha, not numeric. The suffix indicated the engine's intended use. An example of F0112RF translates to a fuel-injected engine (RF), built on January 12th (0112) at the Flint engine plant (F).

The suffix number is one presumed to be correct in a numbers-matching Corvette. Corvette engine suffix numbers have almost always been exclusive to Corvette use. Through 1996, the number containing the suffix was hand-stamped into the engine during assembly by an employe using a tool containing the complete number set. The employe had several number sets to select from and changed sets as engines moved by on the line. Were engines stamped incorrectly? Of course. What happened if the error was recognized? No, the engine wasn't scrapped. Usually, the incorrect number was removed with a grinder, and the correct number stamped over the mistake.

Starting in 1997 when engine production moved to Romulus, codes were no longer hand-stamped during engine assembly. Rather, Romulus attached paper labels containing the suffix (broadcast code) and a barcode. The barcode was read by a special scanner at the vehicle assembly plant and data stored in computer files.

When a Corvette engine is built, the engine plant knows the intended use of the engine, but not into which individual Corvette it will be installed. So when the engine is mated to a vehicle at the Corvette assembly plant, another set of numbers containing the sequential part of the vehicle identification number is stamped into the engine to key it to its specific body and chassis. Starting during 1960 production through 1991, this *vin derivative* was stamped into the same pad as the suffix number. From 1992 through 1996 the location changed to a similar pad at the rear of the engine, usually on the driver's side, but sometimes on the passenger's. The location changed again starting in 1997 to a vertical pad just forward of the bell housing on the driver's side. More than any other number, it is the engine's vin derivative *match* to the vehicle number that constitutes *matching numbers* terminology.

The *Corvette Black Book* also lists a *block* number for engines. This refers to a seven or eight digit block casting number which is cast into the block at the foundry. While a Corvette engine suffix is nearly always for Corvette use only, the same basic block casting may have been used for other Chevrolet or General Motors products. Nevertheless, this is one more way to authenticate a Corvette. The complete casting number for six-cylinder Corvettes is located forward and just below the fuel pump mounting. For V8 engines through 1996, the complete number is located at the upper rear of the driver's side of the block, near the flywheel attachment.

The *head* numbers appearing in the *Corvette Black Book* refer to cylinder head casting numbers. These are normally visible only with the valve covers removed.

Other numbers listed include *carburetor, distributor, generator, alternator*, and *starter*, all engine components. Because of the probability of failure, it isn't unusual for any of these to have been replaced during a Corvette's lifetime. These are also the most difficult for any publication to document accurately and completely. As a general rule, the older the Corvette, the more accurate the component part listing appearing in the *Corvette Black Book*. This is because of the enthusiast interest in older models and the documentation done by enthusiasts and organizations. Later models have had less scrutiny, and the difficulty in seeing component numbers in the crowded engine bays of later Corvettes is another factor in making documentation more difficult.

Another category of numbers shown for some model years is *ending vehicle number*. This is simply the serial number of the last vehicle produced during the last production day of the months listed. This information is not available for all years, and for some years only partial data exists. When a particular month is not listed, it means either that no cars were produced that month, or that the production records for that month are unavailable.

An individual Corvette has scores of date-coded parts not specifically listed in the *Corvette Black Book*. One worthy of explanation here is the block casting date code. This is a date code indicating the date the block was cast in the foundry. Obviously, this does not match either the date of engine manufacture or vehicle manufacture. But it must, with very rare exception, precede engine and vehicle manufacture. Date codes for six-cylinder Corvette engines were on the passenger side of the engine, near the starter solenoid. For most Corvette V8 engines, the date code was at the rear, passenger-side of the block, near the flywheel attachment. Exceptions were 1965 through 1968 and early 1969 big blocks which had their date codes forward of the starter on the passenger side by the freeze plug, 1965 small block #3858180 (cast at Tonawanda but built at Flint) which had its date code adjacent to its casting number, and 1997 and newer aluminum blocks which used a different dating system entirely.

Through 1996, block date codes appeared as three or four characters, an alpha followed by two or three numeric. The alpha denoted the month, "A" for January continuing through "L" for December. The date was indicated by the next one or two numbers; the year by the last single number. A date code of B129 indicates a block cast on February 12, 1959, 1969, 1979 or 1989.

The aluminum blocks used in Corvettes starting with the 1997 model were supplied by two non-GM vendors. One of two suppliers of 1997 and newer blocks (Montupet) used pattern molds and cast a "clock" into the block with months shown around the circumference and year in the center. The other supplier (Nemak) used a metal mold and stamped a dated serial number into a machined pad in the left front. In both cases, a barcoded 16-character engine serial number label breaks down as follows: First two (10) indicate engine, next three indicate broadcast code (suffix), next one blank, next one indicates engine plant (W=Romulus, K=St Catharines, 3=Wixom), next two indicate calendar year, next three indicate Julian date of build (001 is January 1, 365 is December 31st), and last four indicate serial number of engine on that Julian date of build.

Facts

The *facts* section in the *Corvette Black Book* contains details that make each year unique. This section is intended to supplement the hard data presented in the numbers, options, and colors sections. Don't assume a particular Corvette model with a short facts section somehow lacks interest. The size of the facts section is determined by the length of the numbers, options and colors sections, not by the availability of unique features. One reason for including the chronology of all Corvette years is to include a more equally balanced presentation for all years, something not possible in the data section due to the long numbers, option and color listings, especially for years such as 1968, 1969, 2008, and 2009.

Options

Most Chevrolet promotional literature and advertising has contained disclaimers such as: *accurate at press time, but manufacturer reserves the right to change options, colors, and prices without notice.* Things did change, and not just at new model introduction. Corvette prices have changed four times during a model year. Manufacturing problems, within Chevrolet and at suppliers, have caused advertised options not to be available. An option appearing after the start of production might not have appeared in advertising or promotional material.

Understanding the options available during a specific Corvette year, especially what was "rare," is one key to comprehending the hoopla engulfing certain models. The *Corvette Black Book* presents options in a straightforward table, listing the order code (usually Regular Production Option or RPO), a brief description of the option, the quantity sold or produced if known, and the retail price. Further explanations of specific option content, restrictions, or other pertinent data are provided below the tables.

The quantities sold or produced came directly or indirectly from Chevrolet. Chevrolet's method of tracking option quantities has varied over the years, as has its accuracy. Quantities for some years are simply not available and may never be. Some quantities for specific options vary in different Chevrolet records. In the computer age, one would expect counts to be precise, but inconsistencies still exist. It might be tempting to criticize Chevrolet for a lack of precise record keeping over the past decades of Corvette production, but the reality is that the records are quite good considering the complexity and nature of Corvette production when options and option combinations are considered.

Prices listed are *not* current market values. Option costs are retail, or "sticker" prices in effect when these Corvettes were sold new. We use each year's earliest available complete price schedule. Price changes during the year, especially base prices, were uncommon during the early years of Corvette production, common starting in the sixties. Most years since have had one to four price changes. Destination freight charges are generally included in the retail base vehicle prices shown.

Colors

Prior to the 1963 model Corvette, no paint identification coding was affixed to the Corvette body. If an earlier Corvette is repainted carefully, the color can be changed without detection. Easier said than done, and most repaints can be spotted if inspected closely. Still, the lack of positive code verification keeps the color controversy brewing.

The "code" shown in the tables is the exterior color code which appears on a plate or label affixed to Corvette bodies starting in 1963. The plate is on the instrument panel support member under the glovebox for 1963 through 1967 models, on the driver door post for

14

1968 through 1982 models, and on a label for newer models either inside the console lid, glovebox door, or inside one of the behind-seat storage compartments.

Chevrolet used lacquer for the exterior body surfaces of Corvettes built in Flint and St. Louis. The 1957 and earlier Corvettes, except for 1957's Inca Silver and its Imperial Ivory cove, were nitrocellulose lacquer. Inca Silver, its Imperial Ivory cove, and later lacquers were acrylic. The 1981 Corvettes built at Bowling Green and later models have an enamel-type paint with clearcoats.

For refinishing, the newer style paints used by Bowling Green are readily available. Acrylic lacquers are available, but sources are limited. The nitrocellulose colors can be a problem, since many paint formulas for the older colors have been pulled from local suppliers by paint manufacturers. Restorers have the choice of cross-referencing to an acrylic equivalent, or using original nitrocellulose lacquer supplied by vendors servicing the old car market. Be aware that the sale of any lacquer in some areas, including California, is restricted.

The chart paint quantities don't always add correctly due to record-keeping glitches. Wheel colors listed are for standard equipment. Interiors are those recommended by Chevrolet. Up until 1989, it had been Chevrolet's policy to permit customers to override the recommended colors, so different interior-exterior combinations from those shown in the charts are possible. In 1989, Chevrolet removed the color-override option though it still happened in special cases. In 1994, Chevrolet reverted to generally letting customers choose whatever they wanted, but continued to recommend desirable combinations. For 2005 and 2006, non-recommended combinations were not allowed. Starting in 2007, an extra cost option (RPO D30) was added to permit customers to order non-recommended combinations.

Interior color codes, added to Corvette bodies starting in 1963 along with exterior coding, are found under the color charts for each model year, along with other pertinent color data.

Extrapolation

The most common inquiry from readers of the *Corvette Black Book* over the years has centered on the precise number of Corvettes built with a specific combination of options. For example, an owner of a yellow 1979 Corvette with an automatic transmission and trailer package might want to know how many other yellow 1979s shared those features.

Ads for Corvettes and other collectible cars often contain a statement such as "one of only seven built..." so it is not unreasonable to believe that auto companies tabulate all the combinations sold. But they generally don't, because as the option list grows, the number of combinations possible becomes enormous. This is not to say it never happened. If someone at Chevrolet needed to know an option combination, perhaps to evaluate a marketing decision, a manual count of build sheets could have been done, or a program could have been written to search computer files. So when you see a combination-of-options number in print, it may be precise, or it may be an extrapolation.

Extrapolation? A dictionary definition is "to project known data or experience into an area not known or experienced so as to arrive at a conjectural knowledge of the unknown area." Said another way, it means using what you know to guess what you don't.

Using the *Corvette Black Book*, it is possible to extrapolate an answer for the owner of the 1979 Corvette mentioned above. On page 79, the *Corvette Black Book* states that of 53,807 Corvettes sold in 1979, just 2,357 were yellow. In the listing under options, the quantity for automatic transmission quantity is shown as 41,454. That is 77% of the 53,807 total, so the number of yellow cars with automatic transmission can be calculated to be 77% of 2,357, or 1,815. How many of the 1,815 had a trailer package? The trailer package was rare, just 1,001, or 2% of the 53,807 total sold. So an extrapolated estimate of the number of yellow 1979 Corvettes with automatic transmission *and* trailer package is 2% of 1,815, or 36. Remember, this is an estimate, but an educated one. Learn to use extrapolation to zero-in on Corvettes with genuinely rare option combinations.

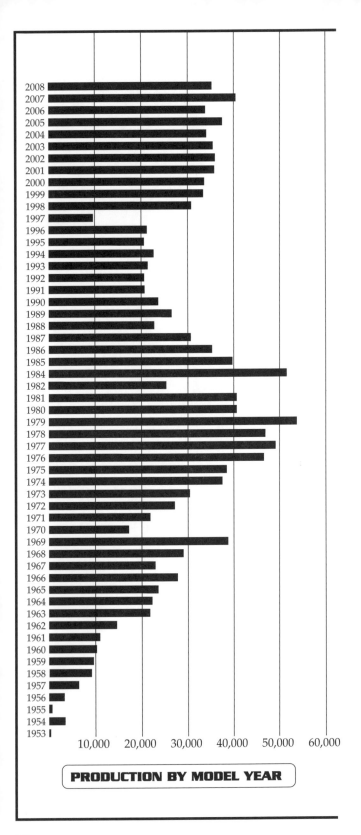

PRODUCTION BY MODEL YEAR

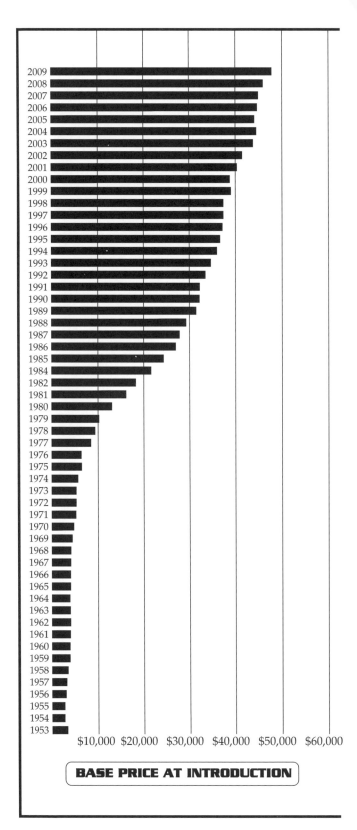

BASE PRICE AT INTRODUCTION

Chronology

1953: The Corvette's debut was in January 1953 at the GM Motorama at the Waldorf Astoria hotel in New York City. The Motorama Corvette was a pre-production prototype and by June 1953 the Corvette was in production at a temporary facility in Flint, Michigan, where 300 Corvettes were built that year. All 1953 models were white with red interiors, and had black canvas tops. All had two-speed Powerglide automatic transmissions mated to 150hp, six-cylinder engines. Although it was anything but a racer, *Speed Age* magazine praised the car in its December 1953 issue in a feature article entitled *Chevrolet Can't Miss With the Corvette*. This was the rarest of all Corvette model years.

1954: Corvette production moved to a renovated facility in St. Louis, Missouri, where production started in December 1953. Chevrolet had projected annual sales of 10,000, but this was optimistic as 3,640 1954 Corvettes were built and nearly a third were unsold at year's end. A mid-year camshaft change increased the six-cylinder engine's horsepower to 155; all 1954s continued with Powerglide automatics. A beige interior and at least three new exterior colors, blue, red and black, were added. Few 1954 models saw race action, but Tom Benavides did drive a modified Corvette to a C-modified second-place finish in the SCCA event at March Air Force Base in California.

1955: Next to 1953, this was the lowest volume Corvette with production of 700 cars. Many predicted this would be the Corvette's last gasp, but Chevy held on, perhaps embarrassed by the success of Ford's 1955 Thunderbird, also a two-seater in its first three years. Most 1955 Corvettes were fitted with a new 265ci, V-8 engine, but a small number of six-cylinder models were built. A three-speed manual transmission was available along with the Powerglide.

1956: The 1956 Corvette body was all-new, and included roll-up glass windows with optional power assists. Interiors were limited to red or beige, but six exterior colors were available. Exteriors could be ordered in two-tones with a contrasting color in the side "cove," a first for Corvette. Engines included a base 210hp V8, a 225hp V8 with dual carburetion, and a dual-carb engine with special camshaft (sometimes rated at 240hp). Seat belts were available for the first time in a Corvette as a dealer-installed option. A removable hardtop was optional. Dr. Dick Thompson and his Corvette won the Sports Car Club of America (SCCA) Class C-production racing title. Sales increased to 3,467.

1957: One of the great automotive milestone models, the 1957 Corvette was the first to combine fuel injection with a four-speed manual transmission. Displacement increased to 283ci, and the most potent fuel-injected engine developed 283hp. Driving a 1957 Corvette, Dr. Dick Thompson and Gaston Audrey won the 12-hour Sebring in March 1957. Corvettes won the SCCA Class B-sports national title (J. E. Rose), and Class B-production Sports (Thompson). Sales rose to 6,339.

1958: Design features included new body panels, new instrument panel with tachometer, and new upholstery. Although criticized by some as overly styled, 1958's distinctive louvered hood and chrome trunk spears set it apart visually from the otherwise similar models which followed. It was the first Corvette with four headlights and factory-installed seat belts. All 1958 model exteriors were painted with acrylic lacquer. Corvettes were virtually unchallenged in SCCA Class B-production with Jim Jeffords taking the points championship. Sales increased to 9,168.

1959: The hood louvers and chrome trunk spears from the previous year were removed and a pure black interior color, the Corvette's first, was available. Jim Jeffords and his Corvette again won the SCCA Class B-production points championship. The top twenty B-production positions were all claimed by Corvettes. Sales increased to 9,670.

1960: The 1960 Corvette was the first with an aluminum radiator (limited to high-lift camshaft engines), the last to have taillights formed into the rounded rear fenders, and the first to exclude the combination of automatic transmission and fuel-injected engine. A Corvette in the Briggs Cunningham team finished eighth overall in the GT class at the 24-hours of Le Mans. Production of 10,261 was the first to reach five figures. Ferrari competed well in SCCA Class B-production but finished second to Bob Johnson and his Corvette.

1961: The 1961 had new rear styling, the first with a four taillight look that became a tradition. It was the last to feature optional two-tone side cove paint treatment and 283ci engines. It was the first with exhaust exits below the body and the first Corvette V8 with radiator expansion tanks. Production edged up to 10,939. In SCCA Class B-production, Dr. Dick Thompson won the points championship with Corvette.

1962: Two-tone paint was no longer available for 1962. This was the last Corvette with a conventional trunk until the 1998 convertible, and the last with a solid rear axle which marked the end of the Corvette's first generation. Engine displacement increased to 327ci. Powerglide transmissions had aluminum cases for weight savings. Sales jumped sharply to 14,531. Corvettes won the SCCA points championships for A-production (Dick Thompson) and B-production (Don Yenko).

1963: The 1963 Corvette featured a new chassis, including independent rear suspension with a transverse leaf spring, and a new body design, available for the first time in both fixed coupe and convertible styles. For this year only, coupe models had a body-color section dividing the rear window. The second genuine Corvette milestone, the 1963 model was a tremendous success with total sales of 21,513. Demand exceeded supply and some buyers waited many months for delivery. Knock-off wheels were optional and appeared in brochures, but probably were available only over-the-counter. Air conditioning and leather seats (saddle only) were available for the first time. Special racing package RPO Z06 initially included a 36-gallon fuel tank and was limited to coupes. Later, Z06 was available with convertibles and standard fuel tanks. Despite the significant Corvette chassis improvements, Carroll Shelby's Cobra also debuted in late 1962 and its superior power-to-weight ratio enabled it to dethrone the Corvette in SCCA racing. Five factory lightweight Corvette racers called Grand Sports were built. All are in private hands today.

1964: The "split" was removed from coupe rear windows to improve visibility, and some styling details of 1963 were deleted including the simulated air vents in the hood and bright instrument trim. Optional knock-off wheels were definitely delivered with 1964 Corvettes to retail customers. Optional leather seating was available in all colors. Power output of the fuel-injected V8 reached 375-hp. Sales continued to improve, climbing to 22,229 and weighted 63% (13,925) toward the convertible body style.

1965: "Big block" engines made their Corvette appearance as RPO L78 with 396ci and 425hp, and models with this engine were fitted with distinctive hoods. It was the last year for first-generation fuel injection introduced in 1957. Four-wheel disc brakes were included in the Corvette's base price, but 316 drum-brake 1965s were built as a delete-cost ($64.50) option while supplies lasted. Options first available included side-mount exhausts, teakwood steering wheels, telescopic steering columns, and goldwall tires. Sales totaled 23,564.

1966: Displacement of optional big block engines increased to 427ci. Power ratings for big blocks were either 390hp or 425/450hp. It was the first year of availability for seat headrests and shoulder harnesses. It was the last for optional knock-off wheels, which in 1966 had brush-finish center cones instead of bright. Seats had additional pleats to reduce seam splitting. Backup lights became standard equipment. Sales were the highest in the Corvette's history up to that time, 27,720.

1967: Last of the 1963-67 "mid year" series and the Corvette's second generation, the 1967 is now an enthusiast favorite. But it was somewhat ignored at the time because buyers anticipated the new body style coming the following year. The 1967 optional aluminum "bolt on" road wheel was unique to the year. The side fender louver treatment was unique, as was the hood for models with optional 427ci engines. Horsepower ratings ranged from the base engine's 300hp to 435hp for the RPO L71 engine. Output of the limited production RPO L88, of which only twenty were sold in 1967, exceeded 500hp but was intentionally understated to restrict its appeal to non-racing customers. A competition-prepared L88 set the A-class Grand Touring land speed record at 192.879 mph at the Bonneville Salt Flats. Emergency brake handles for 1967 moved to between the seats from under the instrument panel, and this was the first Corvette with standard four-way flashers, dual master cylinders, turn signals with lane-change feature, and folding seatback latches. Sales declined to 22,940.

1968: The chassis remained virtually unchanged, but the body and interior for 1968 were completely new, based on the Mako Shark show car. But it wasn't called a Mako Shark as expected, or a Sting Ray as in 1963 through 1967. For the first time, the coupe featured removable roof panels (T-tops) and a removable rear window. Wheel width increased to seven inches. This was the first Corvette without side vent windows and the last (until 1997) with an ignition switch on the instrument panel. To clean up the engine bay and improve weight distribution, the battery was moved to a compartment behind the seats. Despite media criticism of spotty quality—especially from *Car and Driver* magazine which refused to road test a 1968 Corvette due to its poor quality—28,566 of the 1968 models were sold, a new record. Chevrolet dealer and Corvette driver Don Yenko won the SCCA Class A Divisional Championship.

1969: A strike interrupted 1969 production. When it was settled, John DeLorean, Chevrolet's general manager, extended Corvette production by four months. Total production for the 1969 model year was 38,762, a level not reached again until 1976. The 1969 looked similar to 1968, but it did have optional side exhausts, the ignition switch was moved to the steering column, map pockets were added to the instrument panel, and the steering wheel diameter changed from sixteen inches to fifteen. Wheel width increased from seven to eight inches. The Stingray (one word) name reappeared on the car's exterior. Small block engine displacement increased from 327ci to 350ci. Corvette won the GT-class at the Daytona Continental.

1970: Optional big block engine displacement increased to 454ci. The small block remained 350ci, but a new high performance RPO LT1 generated 370hp. This was the last year for high compression engines in this era. The body shape was revised to include fender flares behind the wheels to reduce body damage from wheel-thrown debris. 1970 model build didn't begin until January 1970, and sales fell to 17,316, lowest since 1962. Corvette had GT-class wins at the 12-Hours of Sebring and at the Daytona Continental. Corvette also won SCCA's A-and B-production season championships.

1971: This was the last Corvette to have the fiber-optics light monitoring system introduced with the 1968 model. Compression ratios were lowered in all engines to permit use of lower octane low-lead and unleaded fuels. Production resumed a normal cycle and sales rebounded to 21,801. Dave Heinz won the GTO-class in the inaugural International Motor Sports Association (IMSA) race.

1972: Alarm systems, optional on some earlier Corvettes, were standard for the 1972. This was the last year for a removable rear window in coupes and the last with chrome bumpers front and rear. Power ratings for all engines were reduced due to a change in measurement technique. Instead of "gross," it became a more realistic "net," which included losses from such things as accessories, air cleaners, and mufflers. Sales rose to 27,004. Corvette won the IMSA GT manufacturer's championship.

1973: This model is easily recognized by its body-color front bumper system combined with chrome rear bumpers. The change in front was dictated by federal legislation. The 1973's hood extended over the windshield wipers, eliminating the separate wiper panel used from 1968 to 1972. All 1973 Corvettes had metal beams inside the doors for side impact protection. Corvette led the 12-Hours of Sebring before retiring, and led the Daytona continental to finish 2nd overall. Sales reached 30,464, with 4,943 of these convertibles.

1974: This was the first Corvette with body-color bumpers front and rear. These models were distinguished by the vertical split in their rear bumpers, unique to the year. As the last Corvette model without catalytic converters for emission control, it was the last to legally use leaded fuel (except for export). It was also the last year for availability of a "big block" engine. In racing, Corvette had GT-class wins in IMSA events at Daytona and Talladega, plus SCCA championships in A- and B-production. 37,502 were sold, of which 5,474 were convertibles.

1975: It was the first Corvette model with a one-piece rear bumper skin, catalytic converters, and a pointless (high energy ignition) distributor. At the time—due to pending federal legislation— it was thought this would be the last model year for a convertible, but the convertible body style returned a decade later in 1986. John Greenwood won the SCCA's Trans-American Championship. Sales increased to 38,465, of which 4,629 were convertibles.

1976: The convertible was gone, but that didn't stop 46,558 people from purchasing new 1976 Corvette models. Aluminum wheels were genuinely available after a false start in 1973. Unlike the five-wheel sets of the C2 generation, 1976's wheel option included four aluminum wheels and a steel spare. A steel underpan was added for body rigidity, heat isolation and weight reduction. This was the first Corvette with in-glass heating elements in the rear window defogger, instead of forced air. Corvette won SCCA championships in A- and B-production.

1977: The 1977 was the last Corvette with the vertical rear window treatment started in 1968. Vinyl seating surfaces were banished. Leather was standard, but cloth with leather trim could be substituted at no cost. The console was a new design pulled forward from the 1978 model and accepted a wider variety of radios thanks to its deeper radio recess. Cruise control became available, but only when ordered with automatic transmission. Engine paint color changed during 1977 from Chevrolet Orange to blue. Corvette won SCCA championships in A- and B-production. Sales were 49,213.

1978: Corvettes were big news in 1978, the model's twenty-fifth year. Silver anniversary paint schemes, and "pace car" replicas of the car used to pace the 1978 Indianapolis 500 were offered. The replicas were black over silver and had separate vehicle identification number sequencing. A buying frenzy prior to the race boosted pace car prices to double their $13,653.21 sticker, but this speculative bubble burst following the Indianapolis race. Redesigned elements for all 1978 models included a fixed fastback rear window design, and a new instrument panel with a conventional glovebox. Sales were 46,776, including 15,283 with silver anniversary paint, and 6,502 pace car replicas. Corvettes won the SCCA Trans-Am Series Category II championship as well as the A- and B-production, and B-stock, B-prepared and B-stock ladies Solo II national championships. Also, Corvette won IMSA's AAGT manufacturer's title.

1979: New seat designs, first seen with 1978's pace car model, were standard equipment. Front and rear spoilers, also pace car pieces, were optional in 1979 as RPO D80. Sales soared to 53,807, a record for single-year Corvette sales which still stands today. Even 1984's total, inflated because there was no 1983 model, was less. Corvette won the SCCA Trans-Am Series Category I championship and national titles in B-production, B-stock, B-prepared and B-stock ladies Solo II.

1980: Corvettes had a new look in 1980 thanks to new front and rear bumper caps with integrated spoilers. Federal law required that 1980 cars, Corvettes included, have speedometers with maximum readings of 85 mph. Gulp. The mid-engine GTP Corvette debuted on the IMSA circuit with a 1,200 horsepower V6 and speeds exceeding 200 mph. Double gulp. In an unusual turn of events relating to California emission requirements, Corvette buyers there got a 305ci engine instead of 350ci in 1980, the only year this happened. Sales declined to 40,614.

1981: The St. Louis factory built its last Corvette on August 1, 1981. Bowling Green started production June 1, so for two months both built simultaneously. This was the only time in Corvette history that plant builds overlapped. St. Louis built mostly solid colors using lacquer; Bowling Green built mostly two-tones using a new enamel-type paint with clearcoats. A total of 40,606 were built including 8,995 at Bowling Green. Corvette won the SCCA Trans-Am series championship.

1982: This model ended the third generation (C3) that began in 1968, and the chassis design that dated to 1963. A "collector edition" had silver-beige paint, special trim and a hatchback rear window. The hatch window was Corvette's first, but set off a trend. At $22,537.59, the collector edition was the first Corvette with a base price to exceed $20,000. Manual transmissions were not available. Sales were 25,407 including 6,759 Collector Editions. Corvette was second in the SCCA Trans Am championship and won one race in the IMSA GTO class.

1984: Since this Corvette—the first of the fourth generation— was behind schedule, met 1984 safety and emission standards, and because regular production was scheduled to begin after January 1, 1983, Chevrolet skipped the 1983 model year designation and introduced this mostly-new Corvette as a 1984 in March 1983. Offered only as a coupe with a single "targa" removable roof panel, it had digital instruments, an optional (no-cost) four-speed manual transmission with overdrives in the top three gears, Girlock disc brakes, single front and rear plastic leaf springs mounted transversely, multi-adjustable seating, and extraordinary handling praised by media worldwide. Body seams were hidden behind a center rub strip to eliminate all seam finishing. Corvette won the SCCA showroom stock GT national championship. The extended production run racked up final sales of 51,547.

1985: Ride quality was improved by reducing spring rates for base and RPO Z51 models. Z51s had thicker stabilizers to compensate for the softer springs. New Bosch fuel injection featured tuned runners, a mass airflow sensor and individual injectors for each cylinder. Power increased from 205hp to 230hp with no fuel economy penalty. *Car and Driver* magazine proclaimed the Corvette to be America's fastest production automobile, capable of 150-mph top speed. In racing, Corvette won all six SCCA Showroom Stock endurance races and won the series championship. The Corvette GTP took the pole at Daytona with a new IMSA lap record. Production was 39,729.

1986: Convertible models returned in 1986. Corvette paced the 1986 Indy 500 and all 1986 convertibles were considered replicas, but didn't have special paint or options. Anti-lock brakes (ABS) were standard. Aluminum heads were introduced during 1986. All convertibles had them; early coupes didn't. All 1986 Corvettes had center, high-mount stoplamps. Production of 35,109 included 7,315 convertibles. Corvette won all seven SCCA Showroom Stock endurance races. The Corvette GTP won seven IMSA poles and two races.

1987: The new RPO Z52 combined elements of RPO Z51 with softer base springs. Roller valve lifters reduced engine friction and added 5hp. Callaway twin-turbo conversions were available through some Chevy dealers as RPO B2K. Not a factory option, B2K triggered a special build at Bowling Green before shipment to Callaway Engineering in Old Lyme, Connecticut for engine conversions. Callaway production for 1987 was 188. Total Corvette production was 30,632, including 10,625 convertibles.

1988: A 35th anniversary edition combined white exterior with white leather and black accents. Front suspensions were redesigned to have zero scrub radius steering. Wheel offsets changed and two new wheel designs were introduced. Fifty-six street-legal Corvettes were built for the Corvette Challenge SCCA showroom stock race series. Callaway built 125 twin-turbos. True black—not charcoal—interiors were available. Total Corvette production was 22,789, including 7,407 convertibles.

1989: Bowling Green built 60 cars for the new Corvette Challenge race series, and just over 30 were converted to race use. A new ZF 6-speed manual transmission was introduced. The RPO FX3 adjustable suspension permitted shock absorber firmness change with a console-mounted three-position switch. The much-awaited ZR-1 model, intended for 1989 debut, was delayed to 1990. Callaway sold 67 twin-turbos. This was the last year for the digital instrument panel introduced in 1984. Production totaled 26,412, including 9,749 convertibles.

1990: The big news was introduction of RPO ZR1, available only with coupes. It included special rear body panels and an LT5 engine designed jointly by Lotus and Chevrolet. The engine had an aluminum block and cylinder heads, four overhead camshafts, four valves per cylinder, and was rated at 375hp. LT5 engines were built by Mercury Marine at its Stillwater, Oklahoma factory, but ZR-1s were built on-line at the Bowling Green Corvette plant. All 1990 Corvettes, ZR-1s included, had redesigned interiors (except seats) with a driver-side airbag, and a "hybrid" instrument display with analog tachometer and secondary gauges, and a digital speedometer. Corvette built 23 specially-optioned cars for the SCCA World Challenge series. Callaway sold 58 twin turbos. Total Corvette production was 23,646, including 3,049 ZR-1 models, and 7,630 convertibles.

1991: Base and ZR-1 Corvettes had restyled front-end appearance. The rear of base models was restyled with a convex fascia to look similar to ZR-1 models. The high-mount center stop lamp was mounted in the rear facia for all models except for ZR-1 models where it continued to be roof-mounted. RPO Z51 was not available; it was replaced with RPO Z07 which combined elements of Z51 with the adjustable suspension package. It was available with coupes only. The SCCA World Challenge series continued. No Corvettes were factory-built for the Challenge series, but Corvette did win the manufacturer's championship. Wheel sizes were continued from 1990, but wheel appearance was new. A sensor was added to the oil pan of all models to detect a low-oil condition. Callaway ended twin turbo production with 71 units which included 12 specially-bodied *Speedster* convertibles. Total Corvette production was 20,639, including 2,044 ZR-1s, and 5,672 convertibles.

1992: The appearance of the 1992 was little changed, but there were two significant technical advances. The base engine was extensively redesigned and renamed LT1. It remained 350ci with two valves per cylinder as before, but output increased to 300hp. Also, all 1992 Corvettes had a traction control system called Acceleration Slip Regulation (ASR). It limited acceleration wheel spin with spark retard, throttle closedown and brake intervention. New Goodyear GS-C tires were standard equipment, a Corvette exclusive for the 1992 model year. Corvette won the SCCA World Challenge manufacturer's championship for the third straight year. Production was 20,479, including 502 ZR-1s, and 5,875 convertibles.

1993: A 40th Anniversary option (RPO Z25) had Ruby Red exterior and interior (leather), and special trim. Subtle changes to the base LT1 made it quieter and increased torque from 330 to 340 lb.ft. Cylinder head and valve train changes increased ZR-1 power from 375 to 405hp. A new Passive Keyless Entry system, GM's first, included a special transmitter to lock and unlock doors by proximity. Base models had smaller front wheels (8.5x17) and tires (P255/45ZR17) and larger rear tires (P285/40ZR17). Production total was 21,590, including 448 ZR-1s, and 5,692 convertibles. ZR-1s placed first and second in the Federation Internationale du Sport Automobile (FIA) GT Invitational class in the Sebring 12-hour race.

1994: Interior redesign included passenger-side airbag, new seats and door trim. Leather was standard. ZR-1 got new five-spoke wheels. Convertible rear windows were glass. Air conditioning changed to R-134a refrigerant. Extended Mobility (run flat) tires were optional. Automatic transmission models received electronic controls for improved shift quality, and a brake pedal interlock which required depression of the brake to shift from Park. Production was 23,330, including 448 ZR-1s, and 5,346 convertibles. A Callaway Supernatural Corvette Le Mans, a carbon-fiber bodied racer, won its class pole position for the 24-Hours of Le Mans.

1995: Exteriors had restyled front fender vents. It was ZR-1s last year. Larger brakes, previously part of Z07 and ZR-1, were standard. Production of Indy 500 replicas, Dark Purple over Arctic White convertibles with special trim, was 527. A sensor was added to show auto transmission fluid temperature. Spring rates for base models were reduced. Run flat tires were available; RPO N84 deleted the spare tire. Total production was 20,742, including 448 ZR-1's, and 4,971 convertibles. Callaway Corvette Le Mans models finished second and third in class at the 24-Hours of Le Mans.

1996: Special models marked the C4's end. A Collector Edition in Sebring Silver had ZR-1-style five-spoke wheels painted silver, leather sport seats with "Collector Edition" embroidery and special trim. The Grand Sport was Admiral Blue with white center stripe and red hash marks on the left front fender, plus sport seats with "Grand Sport" embroidery. Its five-spoke wheels were painted black. Grand Sports had LT4 engines with 330-hp. All LT4s had manual transmissions; standard LT1 engines were automatic only. Production was 21,536, including 5,412 Collector Editions and 1,000 Grand Sports. A Callaway Corvette driven by Almo Coppelli won the SCCA World Challenge Championship Sports Division S2-class.

1997: The 1997 was all new with almost no carryover parts. Body, interior, suspension, and engine were new. An aluminum-block, 346ci, 345hp V8 (LS1) drove a rear transaxle with either a GM 4-speed automatic or Borg-Warner 6-speed manual mated to a Getrag limited-slip differential. Available only as a coupe with removable (targa) roof panel, the 1997 had 17" front wheels and 18" rears. Instrument design was reminiscent of 1963 Sting Rays; large, round analog speedometer and tachometer flanked by secondary instruments. Wheelbase increased 8+"and most other dimensions increased slightly; weight dropped about 80 pounds. Production ramped up slowly after a late start, yielding total sales of 9,752.

1998: Convertibles returned. Its separate trunk with outside access was the first for Corvette since 1962. The convertible top had a glass rear window with in-glass defogger. Corvette paced the 1998 Indy 500 and 1,163 Pace Car convertible replicas were sold with Radar Blue exterior, black / yellow interior, and special trim including yellow wheels. Chevy announced a factory-sponsored race program to feature Corvette C5-R. Active Handling, which altered oversteer or understeer through selective brake application, was an interim option. Sales were 31,084, including 11,849 convertibles.

1999: A hardtop model was new. It shared the external trunk design of the convertible, but its roof was fixed. Hardtop colors and options were limited. Twilight Sentinel, Power Tele-steering, and Head-Up Display were new options. Airbags deployed with lower force. Production was 18,078 coupe, 11,161 convertible, and 4,031 hardtop. The factory-sponsored C5-R racer finished second (GT2) in its first race, the Daytona Rolex 24.

2000: The wheel design was new. Initially, it was painted; then a polished version was optional. Later, a painted wheel with slightly thicker spokes was standard, and the first design remained optional with the polished finish. The magnesium wheel option was reduced from $3,000 to $2,000. Active Keyless Entry replaced Passive Keyless Entry. Coupe seals were improved to reduce water entry at doors and rear hatch. Total sales increased to 33,682, split 18,113 coupe, 13,479 convertible, and 2,090 hardtop. The C5-R factory racer won its first class victory in the American Le Mans Series (ALMS) at Ft. Worth, Texas, driven by Ron Fellows and Andy Pilgrim.

2001: A Z06 hardtop model was introduced. Coupe and convertible power increased to 350hp; Z06 had a 385hp (LS6) engine with a 6-speed manual transmission. Z06s also had titanium exhausts, Goodyear F1 Supercar tires (non-run flat) on unique wheels, and trim differences. Z06s had emergency tire inflator kits. Total sales were 35,627, split 15,681 coupe, 14,173 convertible, and 5,773 Z06. Corvette C5-Rs finished first and second in class (8th and 14th overall) of the 24-Hours of Le Mans.

2002: Output of Z06's LS6 engine increased from 385 to 405hp, torque from 385 to 400 lb-ft. The Z06's rear suspension was retuned with different shock absorber valving; brake disc material was upgraded, and wheel construction changed from forged to cast spun for weight savings. Head-up display, previously unavailable with Z06, became standard. Production was 14,760 coupe, 12,710 convertible, and 8,297 Z06 hardtop.

2003: A 50th Anniversary package was optional with coupes and convertibles. It had Anniversary Red "Xirallic" paint, two-tone Shale interior, special trim, and F55 Magnetic Selective Ride. The latter was a new adjustable suspension with faster response times. It was not available with Z06. A 2003 Corvette 50th Anniversary coupe paced the 2002 Indy 500. Production was 12,812 coupe, 14,022 convertible, and 8,635 Z06.

2004: This last of the C5s had optional Commemorative packages for all models to honor the C5R racer's success at LeMans. All commemoratives had LeMans Blue paint and special trim; Coupes and convertibles had Shale interiors and polished 5-spoke wheels. Commemorative Z06s had black interiors, carbon fiber hoods, red and silver C5R-style striping for hood, roof and rear deck, and polished wheels. All Z06s had upgraded suspensions. Production was 16,165 coupe, 12,216 conv, and 5,683 Z06.

2005: The sixth generation (C6) kept C5's front-engine, rear transmission layout, but was nearly all new otherwise. The engine was a new 364ci (6.0L), 400hp small block. The 2005 returned to exposed headlights, ending a Corvette tradition started in 1963. New options for both body styles included DVD-based navigation, OnStar, XM radio, and seat-mounted side impact airbags. Production was 26,728 coupe and 10,644 convertible.

2006: A Z06 model based on the coupe returned. It had an aluminum frame, fixed magnesium-supported top panel, magnesium engine cradle, specific body panels and trim including carbon fiber front fenders and wheelhouses, and a new small-block V8 with 427ci (7.0L). Output for Z06's LS7 was 505hp and 470 lb-ft. A new 6-speed automatic with steering wheel-mounted paddle shifters was optional with coupe and convertible. Production was 16,598 coupe, 11,151 convertible, and 6,272 Z06.

2007: RPO Z51's larger brakes were added to F55 Adjustable Suspension. New options included two-tone seats for coupes and convertibles, and a non-recommended colors override. Steering wheel radio controls were included with uplevel Bose systems. There were two special editions: The Ron Fellows Z06 honored the famed racer. Painted Arctic White, it was the first signed limited edition (399) in Corvette history. Corvette paced the Indy 500 on May 27, 2007. 500 Atomic Orange convertible replicas were sold. Total production of 40,561 was the highest since 1984.

2008: A new LS3 base engine increased from 364ci to 376ci, and from 400hp to 430hp. Optional dual-mode exhaust (NPP) added 6hp. Uplevel equipment groups had leather-wrapped interiors. OnStar and XM were standard. A green E85 Z06 paced the Indy 500, but a second design, black and silver, was offered to the public. 500 were sold. Also, 505 427- Special Edition Z06s were sold. Total 2008 production dipped to 35,310.

2009: ZR1, Corvette's long-rumored, supercharged "supercar" debuted. Power output was 638hp and top speed over 200mph. Base price, including shipping and a $1,700 gas-guzzler tax, cracked six figures at $105,000. Ten wheel designs/finishes were available on Corvette's coupe, convertible, Z06 and ZR1 models, and all had new variable power steering.

1953 Corvette

Production: 300 roadsters

1953 Numbers

Vehicle: E53F001001 through E53F001300

Prefix: LAY: 235ci, 150hp

Block: 3701481: 235ci, 150hp, fd   3835911: 235ci, 150hp, sd

Head: 3836066: 235ci, 150hp

Carburetor: Carter 2066S #3706151: 235ci, 150hp, fd
Carter 2066SA #3706989: 235ci, 150hp, sd

Distributor: 1112314: 235ci, 150hp

Generator: 1102793: 235ci, 150hp

Starter: 1107109: 235ci, 150hp

Ending Vehicle: Jun 53: 001002
Dec 53: 001300

Abbreviations: ci=cubic inch, fd=first design, hp=horsepower, sd=second design.

1953 Facts

• The first two 1953 production Corvettes were completed June 30, 1953, in Flint, Michigan. The 1953 factory wasn't a factory at all, but a temporary pilot assembly facility in the customer delivery garage, an old building on Van Slyke Avenue. By year's end, 300 of the 1953 model Corvettes were built, much by hand, as production processes were developed for assembling the Corvette's revolutionary fiberglass body. As planned, production for 1954 models moved to a renovated St. Louis facility and began in December 1953.

• Records have been located at Chevrolet which show that the first two production Corvettes were sent to Chevrolet Engineering for evaluation at GM's proving grounds. Many engineering modifications and corrections were made to both cars. Portions of the first two production Corvettes may survive.

• A 1969 contest by *Corvette News* magazine to find the oldest Corvette was won by E53F001003, then owned by Californian Ed Thiebaud. This car was purchased by Les Bieri and Howard Kirsch at the Rick Cole Auction Company's classic car auction in Monterey, California in 1987. Research spearheaded by John Amgwert, editor of the National Corvette Restorers Society's *Corvette Restorer* magazine, documented 003's history. It had been reconditioned by Chevrolet, including a new frame, after testing on GM's infamous "Belgian Block" torture track, then released for public sale in California in late 1953 or early 1954. The car has been restored to factory built condition and was purchased at Barrett Jackson's 2006 Scottsdale auction by Chevrolet dealer David Ressler.

• Due to a shortage of Corvette wheel covers, some of the first 1953 Corvettes built were fitted with "dome" wheel covers common to Chevrolet passenger cars. The twenty-seventh Corvette was delivered to its owner with correct Corvette wheel covers, so it is thought perhaps the first twenty-five Corvettes had the temporary dome style. It is possible dealers replaced some of the dome caps with correct covers.

• Standard 1953 Corvette wheel covers (1953-1955) were single-stamped discs with two chrome "spinner" ornaments attached parallel to a central Chevrolet bow tie emblem. A few early examples, thought to be from a vendor test run, had the spinners mounted perpendicular to the emblem.

• Early wheel disc spinners were plated brass forgings, but vendors later changed to plated zinc die castings.

• Wire wheels were never factory options, but some, both real and simulated, were added by dealers.

• Antennas were standard with all 1953 Corvettes and consisted of a mesh screen fiberglassed into the inside surface of the trunk lid.

• Trunk lids of 1953 models generally did not have the moisture absorbent container for the license recess common to 1954 and 1955 models.

1953 Facts

• The 1953 Corvettes all had two interior hood releases, one for each exterior hood latch.

• All 1953s had "short" exhaust extensions.

• The valve cover of the 1953 Corvette was a unique variation of the standard 1953 Chevrolet passenger car cover. The Corvette version was flattened at the forward end for hood clearance. The cover differed from later years by its dual center hold-down bolts. Later covers were held by bolts around the perimeter. The 1953 cover had the words "Blue Flame" on the passenger side and "Special" on the driver side.

• The 1953 brake and fuel lines ran outside the chassis frame. Later models, starting early in 1954 production, ran inboard of the frame.

• Three gas filler door hinges were used. The first, used to about #20, tended to chip paint. The second corrected that, but limited access. A third design, released between #83 and #90, corrected both problems.

• The engine block, head, valve cover, and the intake and exhaust manifolds were painted in blue-green engine enamel.

• Ignition shielding consisted of upper and lower formed metal shields. Both were painted engine color, not plated.

• The 1953 (and early 1954) radiator surge tank was unique. Its surface was smooth. Later units had two stamped radial rigidity bands.

• The 1953 carburetor connecting linkage was a one-piece stamping.

• The location of the fuel filter in the 1953 model was in the fuel line, just forward of the front carburetor.

• Trunk mats for 1953 models were slightly smaller than later years.

• 1953 engine exhaust valves were shorter than later models.

• The road draft tube in most 1953 models had a smooth top surface. A few 1953s and all later units were stamped with an "X" for rigidity.

• The starters in all 1953s used two-field coils.

• Early 1953 Corvettes (up to #175) used a foot-operated windshield washer assembly. Later were vacuum operated.

• Three separate "bullet" style chrome air inlets were used for all 1953s.

1953 Options

CODE	DESCRIPTION	QTY	RETAIL $
2934	Base Corvette Roadster 300		$3,498.00
101A	Heater 300		91.40
102A	AM Radio, signal seeking 300		145.15

• A 235ci, 150hp engine, Powerglide automatic transmission, vinyl interior trim, whitewall tires, windshield washers, and soft top were included in the base price.

• Prices were effective October 16, 1953. Base price included federal excise tax, and $248.00 delivery and handling. Prices did not include local taxes and other dealer charges. Prices were suggested by Chevrolet and some dealers may have charged different prices.

• Although listed as options, all 1953 Corvettes were built with heaters and radios. Customers could not delete these items from orders.

• The 1953 signal-seeking AM radio was the same as the 1954 unit, except 1953 did not have Conelrad national defense markings.

• Auxiliary hardtops were not available for 1953 models as factory options or as Chevrolet-sponsored dealer accessories. However, aftermarket companies manufactured removable hardtops for 1954-1955 Corvettes and some of these were retrofitted to 1953 models.

• The 1953 heater was not a fresh air type; that is, it recirculated interior cockpit air only.

• All 1953 Corvettes were equipped with tube-type whitewall tires.

1953 Colors

EXTERIOR	QTY	SOFT TOP	WHEELS	INTERIOR
Polo White 300		Black	Red	Red

• Interior and exterior combination shown was the only one available. There were no exceptions.

• All 1953 Corvette soft tops were black canvas.

1954 Corvette

Production: 3,640 roadsters

1954 Numbers

Vehicle: E54S001001 through E54S004640

Suffix: YG: 235ci, 150hp, 155hp

Block: 3835911: 235ci, 150hp, 155hp

Head: 3836241: 235ci, 150hp, 155hp

Carburetor: Carter 2066SA #3706989: 235ci, 150hp, 155hp

Distributor: 1112314: 235ci, 150hp, 155hp

Starter: 1107109: 235ci, 150hp
1108035: 235ci, 150hp, 155hp (later four-coil design)

Ending Vehicle: Dec 53: 001014

Abbreviations: ci=cubic inch, hp=horsepower

1954 Facts

• Chevrolet started production of 1954 Corvettes in a renovated St. Louis assembly plant in December 1953. The plant was designed to build 10,000 Corvettes annually. Demand was misjudged, as it would be six years (1960) before Corvette exceeded 10,000 in annual sales.

• The Blue Flame Six engines used in all 1954 Corvettes had power ratings of 150hp or 155hp. The extra 5-horsepower resulted from a camshaft design change made during 1954 production. Externally, the engines looked the same and 150hp decals were used for both, but the more powerful version can be detected by inspecting the camshaft. Later camshafts had three dots between the fifth and sixth inlet cam lobes.

• Early production (about the first 500) 1954 models had two interior hood releases, one to activate each hood latch. Later models had a single interior release to activate both hood latches.

• The window storage bag for the 1954 was color-keyed to the car's interior. The design was more rectangular than the 1953 bag. The 1954 type also had attaching tabs at both ends which permitted it to be held to the trunk division trim panel with bright turnbuckle fasteners.

• The 1954 valve cover was similar to the redesigned 1954 Chevrolet passenger car's cover. Both were attached to the head by stovebolts around the perimeter. The 1954 Corvette's valve cover was either chrome plated or painted engine enamel blue. "Blue Flame" and "150" decals, reading from the passenger side, were affixed to the painted covers.

• Early 1954s had the "bullet" air inlets common to the 1953 model. But 1954s later than #002906 had a dual "pot" apparatus, intended to reduce the possibility of engine fires and to lower noise at full throttle.

• All 1954 Corvettes had six-volt electrical systems.

• Early 1954 Corvettes had short exhaust extensions. Models later than #002523 had longer extensions with built-in deflection baffles. Both styles were originally stainless steel.

• Ignition shielding consisted of upper and lower stamped metal shields, either painted or chromed. Most 1954 Corvettes appear now with both shields painted or both shields chromed, but the factory did not necessarily match the shields on individual cars.

• The 1954 starters had four field coils, except for very early models which had a two-field coil style.

• The road draft tube in 1954 models had an "X" stamped in the top surface for rigidity. Most 1953 tubes were smooth.

• The 1954's radiator surge tank, except for very early, had two stamped radial rigidity bands formed in the tank. All were chrome plated.

• The 1954's brake and fuel lines were routed inboard of the frame members, except for very early models.

• The floor-mounted dimmer switch was relocated slightly inboard early in 1954 production.

• The 1954 carburetor linkage was a fabricated, three-piece link.

1954 Options

CODE	DESCRIPTION	QTY	RETAIL $
2934	Base Corvette Roadster	3,640	$2,774.00
100	Directional Signal	3,640	16.75
101A	Heater	3,640	91.40
102A	AM Radio, signal seeking	3,640	145.15
290B	Whitewall Tires, 6.70x15	3,640	26.90
313M	Powerglide Automatic Transmission	3,640	178.35
420A	Parking Brake Alarm	3,640	5.65
421A	Courtesy Lights	3,640	4.05
422A	Windshield Washer	3,640	11.85

• A 235ci, 150hp (or 155hp) engine, vinyl interior trim, and soft top were included in the base price.

• Prices included federal excise taxes. Local taxes and dealer charges were not included. Prices were effective October 28, 1954. Initial 1954 pricing was the same as 1953. Prices were suggested by Chevrolet and original customer sales records indicate that the actual prices charged by dealers varied both above and below suggested retail.

• By listing the Powerglide automatic transmission as an option, it was implied that a manual transmission was standard equipment. Not true. All 1954 Corvettes had "optional" Powerglide automatic transmissions; in fact, it is nearly certain that all 1954s were built with all options.

• The 1954 signal-seeking AM radio was the same as 1953's, except all but early 1954 radios had 640-kilocycle and 1240-kilocycle Conelrad national defense emergency markings.

• Auxiliary hardtops were not available for 1954 models as factory options or as Chevrolet-sponsored dealer accessories. However, aftermarket companies manufactured removable hardtops for 1954 (and 1953, 1955) Corvettes and some Chevrolet dealers sold them.

• The 1954 heater was not a fresh air type; that is, it recirculated interior cockpit air only.

• All 1954s were built with windshield washer systems. They were vacuum-operated, activated by a button on the windshield wiper switch.

• Tires were changed late in 1954 production from tube-type to tubeless. It is probable that both types were used simultaneously for some time period. Original 1954 tires were U. S. Royal Air Ride, B. G. Goodrich Silvertown, or Firestone Deluxe Champion. All were wide-whitewall type, with whitewall widths varying from 2.5 to 3 inches.

1954 Colors

EXTERIOR	QTY(est)	SOFT TOP	WHEELS	INTERIOR
Polo White	3,230	Beige	Red	Red
Pennant Blue	300	Beige	Red	Beige
Sportsman Red	100	Beige	Red	Red
Black	4	Beige	Red	Red

• Exterior color quantities are not from Chevrolet records. These are estimates based on surveys, theories, and other data. They should not be relied upon as precise.

• Suggested interiors shown. Other combinations were possible.

• Beige carpet heel pads were beige. Some red carpets had red pads, but most were black.

• Interiors and exteriors were not coded to individual cars. The Polo White, Pennant Blue, Sportsman Red, and Black exteriors are those known to have been used during 1954 production.

• Based on original owner reports, there is a reasonable probability that some 1954 Corvettes were painted exterior colors other than Polo White, Pennant Blue, Sportsman Red or Black. Possibilities include, but are not limited to, Metallic Green and Metallic Bronze. Paint vendor documentation confirms the intent to offer additional colors, but production records have not surfaced to positively confirm actual build.

• All 1954 Corvette soft tops were specified to be beige canvas with top bows painted to match, and it is believed all were. However, the owner of a very late 1954 (eight units from the end of production) has reported that based on old home movies, his car appeared to have a white top when purchased new in 1954.

1955 Corvette
Production: 700 roadsters

1955 Numbers

Vehicle: VE55S001001 through VE55S001700
   • For six-cylinder models, "V" is omitted.

Suffix: FG: 265ci, 195hp, at   YG: 235ci, 155hp, at (6-cyl)
   GR: 265ci, 195hp, mt

Block: 3703524: 265ci, 195hp   3835911: 235ci, 155hp (6-cyl)

Head: 3703523: 265ci, 195hp   3836241: 235ci, 155hp (6-cyl)

Distributor: 1110847: 265ci, 195hp, without vacuum advance
   1110855: 265ci, 195hp, with vacuum advance
   1112314: 235ci, 155hp (6-cyl)

Carburetor: Carter 2066SA #3706989: 235ci, 155hp (6-cyl)
   Carter 2218S #3717687: 265ci, 195hp, fd
   Carter 2351S #3724158: 265ci, 195hp, sd

Generator: 1102025: 265ci, 195hp   1102793: 235ci, 155hp (6-cyl)

Starter: 1107627: 265ci, 195hp, fd   1108035: 235ci, 155hp (6-cyl)
   1107645: 265ci, 195hp, sd

Ending Vehicle:

Jan 55: 001027	May 55: 001300	Sep 55: 001599
Feb 55: 001110	Jun 55: 001389	Oct 55: 001634
Mar 55: 001150	Jul 55: 001489	Nov 55: 001688
Apr 55: 001200	Aug 55: 001555	Dec 55: 001700

Abbreviations: at=automatic transmission, ci=cubic inch, fd=first design, hp=horsepower, mt=manual transmission, sd=second design.

1955 Facts

• Outward appearance of the 1955 Corvette nearly duplicated the previous two years, but the big news was under the hood. The new 265ci V8 engine that debuted in 1955 Chevrolet passenger cars also found its way into the Corvette. But not all 1955 Corvettes were V8-powered, as records indicate seven of the six-cylinder models were also built.

• Electrical systems were changed to 12-volt in 1955 Corvette models, except for six-cylinder models which continued to use the 6-volt systems common to 1953 and 1954.

• Corvettes with V8s in 1955 were identified by an enlarged gold "V" attached over the small "v" in the Chevrolet script on both front fenders. Also, the vehicle identification number (vin) for V8 models started with a "V." Six-cylinder models had standard scripts and no "V" in their vins.

• Corvette production of 700 in 1955 was second only to 1953 in low annual volume. Poor public acceptance the previous year resulted in over 1,100 unsold 1954 models at the start of 1955 production. Despite the low production, 1955 remains one of the most mysterious Corvette models in terms of accurate documentation.

• Tachometers for 1955 models with V8 engines read to 6000rpm in 1000rpm increments. Tachometers for six-cylinder models read to 5000rpm in 500rpm increments (as in 1953-1954).

• Ignition shielding for 1955 consisted of chrome distributor and coil covers with bails, braided and grounded plug wires, and wire carriers behind the exhaust manifolds.

• A manual heater cutoff valve was spliced into the upper heater hose along the inner fender of V8 models.

• Windshield washer activation was by floor pedal with coordinator.

• Valve covers for V8 models were chrome plated with Chevrolet script. They were held in place by Phillips-head screws. The six-cylinder model valve covers were the same as used in 1954.

• Very early in 1955 production, a new style inside rearview mirror was used which permitted vertical adjustment of the mirror base with a thumbscrew mechanism. This mirror also had a slightly larger head compared to the earlier type, aiding rear vision with the top up.

1955 Options

CODE	DESCRIPTION	QTY	RETAIL $
2934-6	Base Corvette Roadster, six-cylinder	7	$2,774.00
2934-8	Base Corvette Roadster, V8	693	2,909.00
100	Directional Signal	700	16.75
101	Heater	700	91.40
102A	AM Radio, signal seeking	700	145.15
290B	Whitewall Tires, 6.70x15	—	26.90
313	Powerglide Automatic Transmission	—	178.35
420A	Parking Brake Alarm	700	5.65
421A	Courtesy Lights	700	4.05
422A	Windshield Washers	700	11.85

• A 235ci, 155hp six-cylinder engine, 3-speed manual transmission, vinyl interior trim, and a soft top were included in the base price of #2934-6. However, the Powerglide automatic transmission was a required option and no 1955 Corvette with the combination of six-cylinder and manual transmission has ever been documented.

• A 265ci, 195hp V8 engine, 3-speed manual transmission, vinyl interior trim, and a soft top were included in the base price of #2934-8. The Powerglide automatic transmission was a required option with the V8 engine until somewhere past the midpoint of 1955 production when the manual transmission started to be used.

• Most 1955 models had automatic transmissions. Estimates place the number of manual transmissions at 75. Though not precise, available records do support a total in the range of 70 to 80 units.

• It is likely most 1955 options were not really optional, but required. Exceptions may surface, but the list of probable 100% usage options includes directional signals, radios, parking brake alarms, courtesy lamps and windshield washers.

• The 1955 heater was a non-fresh air unit; that is, it recirculated interior cabin air only. The heater itself was the same for six-cylinder and V8 models, except for modifications in the blower motors required by the different voltages of the two models.

• Auxiliary hardtops were not available for 1955 models as factory options or as Chevrolet-sponsored dealer accessories. However, aftermarket companies manufactured removable hardtops for 1955 (and 1953-54) Corvettes and some Chevrolet dealers sold them.

• Corvette tires changed from tube-type to tubeless during 1954, so it is likely, but not certain, that all 1955 models had tubeless tires. Both whitewall and blackwall styles were used. Original tires were U. S. Royal Air Ride, B. F. Goodrich Silvertown, or Firestone Deluxe Champion.

1955 Colors

CODE	EXTERIOR	QTY(est)	SOFT TOP	WHEELS	INTERIOR
567	Polo White	325	White/Beige	Red	Red
570	Pennant Blue	45	Beige	Red	Dark Beige
573	Corvette Copper	15	White	Bronze	Dark Beige
596	Gypsy Red	180	White/Beige	Red	Light Beige
632	Harvest Gold	120	Dark Green	Yellow	Yellow

• Exterior color quantities are not from Chevrolet records. These are estimates based on surveys, theories, and other data. They should not be relied upon as precise.

• Interiors and exteriors were not coded to individual cars. Only 700 were produced, yet no Chevrolet records have been found to document color usage. The exterior colors are subject to question and conjecture. Records do show Pennant Blue was discontinued in April 1955. Gypsy Red and Corvette Copper are thought to have been offered after Pennant Blue was discontinued. Owners report other color combinations.

• Yellow interiors came with green carpeting which had a looser weave than other colors. All heel pads were color matched to the carpet color, except red carpets which had either black or red pads

• Early 1955 soft tops were made of a canvas material. A vinyl-coated fabric material was introduced after production started. Both materials were used for beige and green soft tops, but all white soft tops were vinyl. Owner surveys have not determined an exact transition from one soft top material to another. Concurrent use was likely.

1956 Corvette

Production: 3,467 convertibles

1956 Numbers

Vehicle: E56S001001 through E56S004467

Suffix: FG: 265ci, 225hp, at   GU: 265ci, 240hp, mt
       FK: 265ci, 210hp, at   GV: 265ci, 210hp, mt
       GR: 265ci, 225hp, mt

Block: 3720991: All

Head: 3725306: 265ci, 210hp, 225hp (2-bolt exhaust manifolds)
       3731762: 265ci, 225hp, 240hp (3-bolt exhaust manifolds)

Carburetor: Carter 2362S #3720953: 265ci, 225hp, 240hp, rc
             Carter 2366SA #3733246: 265ci, 210hp
             Carter 2419S #3730599: 265ci, 225hp, 240hp, fc

Distributor: 1110872: 265ci, 225hp, ep   1110879: 265ci, 225hp, lp

Generator: 1102043: All

Abbreviations: at=automatic transmission, ci=cubic inch, ep=early production, fc=front carburetor, hp=horsepower, lp=late production, mt=manual transmission, rc=rear carburetor.

1956 Facts

• The Corvette received its first major body redesign in 1956. With the exception of the instrument panel which was nearly identical, the 1956 was visually completely different from the preceding model. The new design featured roll-up glass windows with power assist optional, external door handles and locks, exposed headlights with chrome surround bezels (except very early production which were painted), and a sculptured side cove which permitted the Corvette's first two-tone paint combinations.

• Factory auxiliary hardtops were first available in 1956. The anodized header trim made the 1956 tops unique to the year. The anodized trim was, however, painted the same color as the top on early 1956s.

• Seat belts were available for the first time in 1956, in kit form as a dealer installed accessory (mounting provisions were provided by the factory). The belts were grey nylon with chrome-plated, quick-release buckles.

• The optional AM radios for 1956 Corvettes were transistorized, another Corvette first. A similar radio was used for 1957, but most 1956 selector bars for the signal-seeking feature were plain. Models not equipped with the radio received a painted block-out cover plate.

• All 1956 models with standard transmissions were fitted with a new type clutch which used heat-treated coils to replace the diaphragm-type springs previously used.

• Early production generally had white plastic shift knobs; later production generally had chrome-plated steel shift knobs.

• Dual four-barrel carburetors were available on Corvettes for the first time in 1956. The proper air cleaners were buffed aluminum and had oil-wetted, sealed filter elements. A cast aluminum intake manifold was included with dual carburetors.

• Dual point distributors were first used in 1956. All engines except the base 265ci, 210hp unit had the new distributors.

• The windshield washer reservoir for 1956 was a blue bag made of vinyl and vinyl-impregnated cloth. It was partially transparent to permit viewing of the fluid level. It attached to the left inner fender.

• Correct 1956 valve covers had staggered hold-down holes and attach with Phillips-head screws. The base 265ci, 210hp engine had painted steel covers with Chevrolet script. All optional engines had nine-fin cast alloy valve covers. Engine blocks were painted red.

• Passenger-side seats in 1956 models had fore-aft adjustment rails. Previous Corvette passenger seats were fixed in position.

• A higher-output 12V battery (53-amp hour) was fitted to 1956 models.

1956 Options

CODE	DESCRIPTION	QTY	RETAIL $
2934	Base Corvette Convertible	3,467	$3,120.00
101	Heater	—	123.65
102	AM Radio, signal seeking	2,717	198.90
107	Parking Brake Alarm	2,685	5.40
108	Courtesy Lights	2,775	8.65
109	Windshield Washers	2,815	11.85
290	Whitewall Tires, 6.70x15	—	32.30
313	Powerglide Automatic Transmission	—	188.50
419	Auxiliary Hardtop	2,076	215.20
426	Power Windows	547	64.60
440	Two-Tone Paint Combination	1,259	19.40
449	Special High-Lift Camshaft	111	188.30
469	265ci, 225hp Engine (2x4 carburetors)	3,080	172.20
471	Rear Axle, 3.27:1 ratio	—	0.00
473	Power Operated Folding Top	2,682	107.60

• A 265ci, 210hp engine, 3-speed manual transmission, vinyl interior trim, and a soft top were included in the base price.

• RPO 449 (Special High-Lift Camshaft) was available only when combined with RPO 469 (the 225-hp engine with dual four-barrel carburetors). Chevrolet recommended this combination "for racing purposes only" and generally did not specify a horsepower rating. The accepted, though unofficial, power output of RPO 449 was 240hp.

• RPO 469 was often specified by Chevrolet as "dual four-barrel carburetor equipment," rather than as a separate engine. Of the 3,080 RPO-469 quantity, 1,510 were sold with 3-speed manual transmissions, 1,570 with Powerglide automatics.

• Although the RPO-419 auxiliary hardtop is shown as an option, it could be substituted at no cost for the convertible top. The 2,076 quantity shown included 629 such substitutions.

• RPO 470 permitted substitution of a beige or white soft top in place of the standard black soft top at no charge.

• The Corvette heater was changed from a recirculating-air-only type to a new "outside" fresh air design in 1956. However, the first 145 1956s were built with the old type heater.

• Several of the options shown, including Parking Brake Alarm, Courtesy Lights, Windshield Washers, RPO 469 engines, and Power Top were mandatory purchase early in the production year.

1956 Colors

EXTERIOR	QTY	SOFT TOP	WHEELS	INTERIOR
Onyx Black	810	Bk-W	Black	Red
Aztec Copper	402	Bg-W	Copper	Beige
Cascade Green	290	Bg-W	Green	Beige
Arctic Blue	390	Bg-W	Blue	Beige-Red
Venetian Red	1,043	Bg-W	Red	Red
Polo White	532	Bk-W	Red	Red

• Interiors and exteriors were not coded to individual cars.

• In 1956 paperwork, code "std" denoted black exterior. All other exteriors, including two-tones, were denoted by "440" followed by one alpha character. Solid colors were denoted by letters R through X (no U). Two-tones were denoted by the letters A through G.

• Exterior color quantity total for 1956 from Chevrolet records equals the 3,467 production total exactly. It appears unlikely that any other exterior colors were possible in 1956.

• The 1,259 quantity for code 440 two-tone paint was split 223 Onyx Black/silver, 166 Aztec Copper/beige, 147 Cascade Green/beige, 172 Arctic Blue/silver, 431 Venetian Red/beige, 120 Polo White/silver.

• The silver color used for cove areas was not Inca Silver. Although incorrect repaints are common, the silver used for coves was called Metallic Silver. This was also true for 1957 coves.

• Interiors sold in 1956 were: 2,580 red, 887 beige.

• Headliners were black for black or white tops, beige for beige tops. In 1956, soft top colors sold were 103 black, 1,840 white, and 895 beige.

Abbreviations: Bg=Beige, Bk=Black, W=White.

1957 Corvette

Production: 6,339 convertibles

1957 Numbers

Vehicle: E57S100001 through E57S106339

Suffix:
EF: 283ci, 220hp, mt
EG: 283ci, 270hp, mt
EH: 283ci, 245hp, mt
EL: 283ci, 283hp, mt
EM: 283ci, 250hp, mt

EN: 283ci, 283hp, mt, ai, uu
FG: 283ci, 245hp, at
FH: 283ci, 220hp, at
FK: 283ci, 250hp, at

Block: 3731548: All

Head: 3740997: 283ci, 220hp, 245hp, 250hp, 270hp
3731539: 283ci, 283hp

Carburetor:
Carter 2362S #3720953: 283ci, 245hp, rc
Carter 2366SA #3733246: 283ci, 220hp, fd
Carter 2419S #3730599: 283ci, 245hp, fc, fd
Carter 2613S #3741089: 283ci, 270hp, fc
Carter 2614S #3741090: 283ci, 270hp, rc
Carter 2626S #3744002: 283ci, 245hp, fc, sd
Carter 2627S #3744004: 283ci, 245hp, rc, sd
Carter 2655S #3744925: 283ci, 220hp, sd

Fuel Injection:
Rochester 7014360: 283ci, 250hp, 283hp, ep
Rochester 7014520: 283ci, 250hp, 283hp
Rochester 7014800: 283ci, 250hp, 283hp, lp
Rochester 7014960: 283ci, 283hp, lp

Distributor:
1110889: 283ci, 250hp, 283hp
1110891: 283ci, 220hp, 245hp, 270hp
1110905: 283ci, 250hp(mt,at), 283hp
1110906: 283ci, 250hp, at
1110908: 283ci, 283hp

Generator: 1102043: All

Ending Vehicle:
Oct 56: 100580
Nov 56: 101070
May 57: 104331

Jun 57: 104924
Jul 57: 105584
Aug 57: 106229

Sep 57: 106339

Abbreviations: ai=cold air induction, at=automatic transmission, ci=cubic inch, fc=front carburetor, fd=first design, ep=early production, hp=horsepower, lp=late production, mt=manual transmission, rc=rear carburetor, sd=second design, uu=uncertain usage.

1957 Facts

• Although the body for 1957 was a carryover from 1956, the 1957 Corvette was a milestone model due to its optional fuel injection and optional 4-speed manual transmission, both Corvette firsts.

• Fuel injection was available throughout 1957 production, though limited early. A total of 1,040 fuel injected 1957 Corvettes were sold.

• The 4-speed manual transmission was released for sale on April 9, 1957, equating to about vehicle #E57S103500. Sales were 664 for 1957.

• The optional 1957 AM radio was similar to the 1956 transistorized unit, except most 1957 selector bars contained the word "Wonderbar."

• Both white plastic and chrome-plated steel shift knobs were used. Chrome is generally seen on early cars, white on later.

• Correct 1957 valve covers had staggered hold-down holes and attached with Phillips-head screws. The base 283ci, 220hp engine had painted steel valve covers with Chevrolet script. All optional had seven- or nine-fin cast alloy valve covers. Engines were painted Chevrolet orange.

• Seat belts were not factory-installed in 1957 Corvettes, but mounting provisions were provided for installation by dealers or owners.

• The optional windshield washer reservoir was supplied by Trico. It was a hard, white plastic jar with a red lid.

1957 Options

CODE	DESCRIPTION	QTY	RETAIL $
2934	Base Corvette Convertible	6,339	$3,176.32
101	Heater	5,373	118.40
102	AM Radio, signal seeking	3,635	199.10
107	Parking Brake Alarm	1,873	5.40
108	Courtesy Lights	2,489	8.65
109	Windshield Washers	2,555	11.85
276	Wheels, 15x5.5 (5)	51	15.10
290	Whitewall Tires, 6.70x15	5,019	31.60
303	3-Speed Manual Transmission, close ratio	4,282	0.00
313	Powerglide Automatic Transmission	1,393	188.30
419	Auxiliary Hardtop	4,055	215.20
426	Power Windows	379	59.20
440	Two-Tone Paint Combination	2,794	19.40
469A	283ci, 245hp Engine (2x4 carburetors)	2,045	150.65
469C	283ci, 270hp Engine (2x4 carburetors)	1,621	182.95
473	Power Operated Folding Top	1,336	139.90
579A	283ci, 250hp Engine (fuel injection)	182	484.20
579B	283ci, 283hp Engine (fuel injection)	713	484.20
579C	283ci, 250hp Engine (fuel injection)	102	484.20
579E	283ci, 283hp Engine (fuel injection)	43	726.30
677	Positraction Rear Axle, 3.70:1	327	48.45
678	Positraction Rear Axle, 4.11:1	1,772	48.45
679	Positraction Rear Axle, 4.56:1	—	48.45
684	Heavy Duty Racing Suspension	51	780.10
685	4-Speed Manual Transmission	664	188.30

• A 283ci, 220hp, engine, 3-speed manual transmission, vinyl interior trim, and a soft top were included in the base price.

• RPO 684 included special front and rear springs and shock absorbers, heavier front stabilizer bar, quick steering adaptor, metallic brake facings, finned brake drums, fresh air ducting to rear brakes, and front brake air scoops (elephant ears). Positraction, manual transmission and 270hp or 283hp engines, were required. A version of this option, RPO 581, may have been available earlier without the brake components.

• RPO 276 included wheels which were one-half inch wider than stock, and small passenger-car hubcaps in place of standard full wheel discs.

• The 4,055 RPO 419 (auxiliary hardtop) quantity included 931 in place of soft tops at no charge.

• RPO 579E included fresh air intake and a mechanical tachometer mounted on the steering column.

• RPOs 287 and 288 for 6-ply blackwall tires were planned but actual delivery to retail customers is uncertain.

1957 Colors

EXTERIOR	QTY	SOFT TOP	WHEELS	INTERIOR
Onyx Black	2,189	Bg-Bk-W	Black	Bg-R
Aztec Copper	452	Bg-W	Copper	Bg
Cascade Green	550	Bg-Bk-W	Green	Bg
Arctic Blue	487	Bg-Bk-W	Blue	Bg-R
Venetian Red	1,320	Bg-Bk-W	Red	Bg-R
Polo White	1,273	Bg-Bk-W	Red-Silver	Bg-R
Inca Silver	65	Bk-W	Silver	Bg-R

• Suggested interiors shown. Other combinations were possible.

• Interiors and exteriors were not coded to individual cars. In 1957, three Corvettes were painted a non-standard color, color combination, or left in primer.

• The 2,794 quantity for code 440 two-tone paint was split 10 Inca Silver/ivory, 909 Onyx Black/silver, 319 Polo White/silver, 716 Venetian Red/beige, 263 Aztec Copper/beige, 258 Arctic Blue/silver, 319 Cascade Green/beige.

• Interiors sold in 1957 were: 5,021 red, 1,315 beige, 3 uncertain.

• Wheel color with Polo White exteriors depended on interior color. Wheels were red with red interiors; silver with beige interiors.

• Headliners were black for black or white tops, beige for beige tops. In 1957, soft top colors sold were 1,820 black, 2,461 white, 1,127 beige.

Abbreviations: Bg=Beige, Bk=Black, R=Red, W=White.

1958 Corvette

Production: 9,168 convertibles

1958 Numbers

Vehicle: J58S100001 through J58S109168

Suffix:
CQ: 283ci, 230hp, mt	CU: 283ci, 270hp, mt
CR: 283ci, 250hp, mt	DG: 283ci, 230hp, at
CS: 283ci, 290hp, mt	DH: 283ci, 250hp, at
CT: 283ci, 245hp, mt	DJ: 283ci, 245hp, at

Block: 3737739 or 3756519: All

Head: 3748770: All (sh)

Carburetor: Carter 2613S #3741089: 283ci, 270hp, fc
Carter 2614S #3741090: 283ci, 270hp, rc
Carter 2626S #3744002: 283ci, 245hp, fc
Carter 2627S #3744004: 283ci, 245hp, rc
Carter 2669S #3746384: 283ci, 230hp

Fuel Injection: Rochester 7014800: 283ci, 250hp
Rochester 7014800R: 283ci, 290hp, ep
Rochester 7014900: 283ci, 250hp
Rochester 7014900R: 283ci, 290hp
Rochester 7014960: 283ci, 290hp
Rochester 7017200: 283ci, 250hp, lp

Distributor: 1110890: 283ci, 230hp
1110891: 283ci, 245hp, 270hp
1110908: 283ci, 290hp, ep
1110914: 283ci, 290hp
1110915: 283ci, 250hp

Generator: 1102043: 283ci, 230hp, 245hp, 250hp, 270hp
1102059: 283ci, 290hp

Ending Vehicle:
Oct 57: 100486	Feb 58: 104789	Jun 58: 108192
Nov 57: 101443	Mar 58: 105779	Jul 58: 108840
Dec 57: 102511	Apr 58: 106544	Aug 58: 109168
Jan 58: 103677	May 58: 107489	

Abbreviations: at=automatic transmission, ci=cubic inch, fc=front carburetor, fd=first design, ep=early production, hp=horsepower, lp=late production, mt=manual transmission, rc=rear carburetor, sd=second design, sh=staggered valve cover holes.

1958 Facts

• Extensive redesign for 1958 included new body panels, new instrument panel and new upholstery. External distinguishing features included dual headlights, a Corvette first, nonfunctional louvers on the hood, and twin chrome trunk spears.

• The interior for 1958 had a large 160-mph speedometer flanked by secondary instruments. The tachometer was relocated from its previous central instrument panel location to just above the steering column. The tachometer read to 6000rpm in 1000rpm increments for 230hp , 245hp and 250hp engines. It read to 8000rpm for 270hp and 290hp engines.

• A passenger grab bar was built into the passenger side. There was no package tray. An anodized aluminum trim insert on the passenger side had "Corvette" in white letters. A central console was included. Interior door panels were a two-piece design.

• Seat belts were factory-installed for the first time in 1958 Corvettes. Previously, they had been dealer-installed accessories. However, seat belt anchor provisions were provided for 1956-57 models by the factory.

• Correct 1958 valve covers had staggered hold-down holes and attach with Phillips-head screws. The base 283ci, 230hp engine had painted steel covers. All optional engines had seven-fin cast alloy valve covers.

• Low-loop rayon pile carpeting was factory-installed in 1958 Corvettes.

• Carbureted engines were equipped with glass fuel filter bowls.

• All 1958 exterior paint was acrylic lacquer. Paint for earlier years was nitrocellulose lacquer except for Inca Silver which was also acrylic.

• 1958's grill had nine teeth, instead of thirteen as in previous years.

1958 Options

CODE	DESCRIPTION	QTY	RETAIL $
867	Base Corvette Convertible	9,168	$3,591.00
101	Heater	8,014	96.85
102	AM Radio, signal seeking	6,142	144.45
107	Parking Brake Alarm	2,883	5.40
108	Courtesy Light	4,600	6.50
109	Windshield Washers	3,834	16.15
276	Wheels, 15x5.5 (5)	404	0.00
290	Whitewall Tires, 6.70x15	7,428	31.55
313	Powerglide Automatic Transmission	2,057	188.30
419	Auxiliary Hardtop	5,607	215.20
426	Power Windows	649	59.20
440	Two-Tone Exterior Paint	3,422	16.15
469	283ci, 245hp Engine (2x4 carburetors)	2,436	150.65
469C	283ci, 270hp Engine (2x4 carburetors)	978	182.95
473	Power Operated Folding Top	1,090	139.90
579	283ci, 250hp Engine (fuel injection)	504	484.20
579D	283ci, 290hp Engine (fuel injection)	1,007	484.20
677	Positraction Rear Axle, 3.70:1	1,123	48.45
678	Positraction Rear Axle, 4.11:1	2,518	48.45
679	Positraction Rear Axle, 4.56:1	370	48.45
684	Heavy Duty Brakes and Suspension	144	780.10
685	4-Speed Manual Transmission	3,764	215.20

• A 283ci, 230hp engine, 3-speed manual transmission, vinyl interior trim, and a soft top were included in the base price.

• RPOs 469C and 579D required three-or four-speed manual transmission.

• RPO 684 included special front and rear springs and shock absorbers, heavier front stabilizer bar, quick steering adaptor, metallic brakes, finned brake drums, fresh air ducting to rear brakes and front brake air deflectors (except very early models). Positraction axle, manual transmission and RPO 579D were required.

• RPO 276 included wheels one-half inch wider than stock, and small passenger-car hubcaps in place of standard full wheel discs.

• The 5,607 RPO 419 (auxiliary hardtop) quantity included 2,215 in place of soft tops at no charge.

• The 504 RPO 579 (250hp engine) quantity was split 400 with manual transmissions, 104 with Powerglide automatic transmissions.

• The 2,436 RPO 469 (245hp engine) quantity was split 1,897 with manual transmission, 539 with Powerglide automatic transmissions.

• RPOs 677, 678 and 679 (Positraction) required manual transmission. RPO 679 was split 75 with 3-speed transmission, 295 with 4-speed.

1958 Colors

EXTERIOR	QTY	SOFT TOP	WHEELS	INTERIOR
Charcoal	1,631	Bk-W	Silver	Bg-C-R
Snowcrest White	2,477	Bk-W-Bg	Silver	Bg-C-R
Silver Blue	2,006	Bg-W	Silver	Bg-C
Regal Turquoise	510	Bk-W	Silver	C
Panama Yellow	455	Bk-W	Silver	C
Signet Red	1,399	Bk-W	Silver	C-R
Tuxedo Black	493	Bk-W	Silver	C-R
Inca Silver	193	Bk-W	Silver	C-R

• Suggested interiors shown. Other combinations were possible.

• Interiors and exteriors were not coded to individual cars. In 1958, four Corvettes were painted a nonstandard color, color combination, or left in primer.

• The 3,422 quantity for code 440 two-tone paint (contrasting cove) was split 757 Silver Blue/silver, 756 Signet Red/white, 729 Charcoal/silver, 499 Snowcrest White/silver, 252 Regal Turquoise/white, 190 Panama Yellow/white, 199 Tuxedo Black/silver, 36 Inca Silver/black.

• The Charcoal exterior color is believed to have been replaced by Black somewhere past the midpoint of 1958 production.

• The wheels of RPO 276 may have been painted black instead of silver.

• Headliners for soft tops were black. In 1958, soft top colors sold were 1,864 black, 3,827 white, 795 blue/gray.

Abbreviations: Bg=Blue-gray, Bk=Black, C=Charcoal, R=Red, W=White.

1959 Corvette

Production: 9,670 convertibles

1959 Numbers

Vehicle: J59S100001 through J59S109670

Suffix:
CQ: 283ci, 230hp, mt CU: 283ci, 270hp, mt
CR: 283ci, 250hp, mt DG: 283ci, 230hp, at
CS: 283ci, 290hp, mt DH: 283ci, 250hp, at
CT: 283ci, 245hp, mt DJ: 283ci, 245hp, at

Block: 3737739: All (ep)   3756519: All

Head: 3748770: All (ep, sh)   3767465: All (lp)
3755550: All (ep, sh)

Carburetor: Carter 2613S #3741089: 283ci, 270hp, fc
Carter 2614S #3741090: 283ci, 270hp, rc
Carter 2626S #3744002: 283ci, 245hp, fc
Carter 2627S #3744004: 283ci, 245hp, rc
Carter 2818S #3756676: 283ci, 230hp

Fuel Injection: Rochester 7014900: 283ci, 250hp
Rochester 7014900R: 283ci, 290hp
Rochester 7017200: 283ci, 250hp
Rochester 7017250: 283ci, 290hp
Rochester 7017300: 283ci, 290hp
Rochester 7017300R: 283ci, 250hp
Rochester 7017310: 283ci, 250hp

Distributor: 1110891: 283ci, 245hp, 270hp   1110915: 283ci, 250hp
1110914: 283ci, 290hp   1110946: 283ci, 230hp

Generator: 1102043: 283ci, 230hp, 245hp, 250hp, 270hp
1102059: 283ci, 290hp, fd
1102173: 283ci, 290hp, sd

Ending Vehicle:
Sep 58: 100409   Jan 59: 103962   May 59: 107934
Oct 58: 100623   Feb 59: 104921   Jun 59: 108702
Nov 58: 101587   Mar 59: 106033   Jul 59: 109437
Dec 58: 102641   Apr 59: 107144   Aug 59: 109670

Abbreviations: at=automatic transmission, ci=cubic inch, fc=front carburetor, fd=first design, ep=early production, hp=horsepower, lp=late production, mt=manual transmission, rc=rear carburetor, sd=second design, sh=staggered valve cover holes.

1959 Facts

• Exterior 1959 appearance was similar to 1958, except 1959 did not have the simulated hood louvers or the twin chrome trunk spears.

• Door panels were redesigned by relocating the armrests for additional elbow room, and by moving the door releases forward. Also, the main section of the door panel was a one piece design.

• Instruments were redesigned for better legibility in 1959. This included making the gauge lenses concave for less light reflection, and changing the design of the tachometer face. All 1959 tachometers read to 7000rpm in 1000rpm increments (hash marks at 100rpm increments) with safe, caution and danger zones indicated by pale green, yellow and red.

• The "T" shift handle with positive reverse lockout was introduced in 1959 models with 4-speed manual transmissions.

• A storage bin was added under the passenger grab bar. The grab bar itself was more heavily padded than for the previous year. The anodized aluminum trim insert on the passenger side had "Corvette" in white letters early in production. The letters were changed to black for the balance of production, and the black treatment was more common.

• The optional 1959 windshield washer reservoir mounted on the left side for all carbureted engines, and on the right side for all fuel injected engines. Right side mountings were protected by heat shields.

• Seat upholstery was smoother than the previous year and the black interior color available in 1959 models was a first for Corvette. Seat pleats run side to side in 1959, unlike 1958 and 1960 which run fore and aft.

1959 Options

CODE	DESCRIPTION	QTY	RETAIL $
867	Base Corvette Convertible	9,670	$3,875.00
101	Heater	8,909	102.25
102	AM Radio, signal seeking	7,001	149.80
107	Parking Brake Alarm	3,601	5.40
108	Courtesy Light	3,601	6.50
109	Windshield Washers	7,929	16.15
121	Radiator Fan Clutch	67	21.55
261	Sunshades	3,722	10.80
276	Wheels, 15x5.5 (5)	214	0.00
290	Whitewall Tires, 6.70x15	8,173	31.55
313	Powerglide Automatic Transmission	1,878	199.10
419	Auxiliary Hardtop	5,481	236.75
426	Power Windows	587	59.20
440	Two-Tone Exterior Paint	2,931	16.15
469	283ci, 245hp Engine (2x4 carburetors)	1,417	150.65
469C	283ci, 270hp Engine (2x4 carburetors)	1,846	182.95
473	Power Operated Folding Top	661	139.90
579	283ci, 250hp Engine (fuel injection)	175	484.20
579D	283ci, 290hp Engine (fuel injection)	745	484.20
675	Positraction Rear Axle	4,170	48.45
684	Heavy Duty Brakes and Suspension	142	425.05
685	4-Speed Manual Transmission	4,175	188.30
686	Metallic Brakes	333	26.90
1408	Blackwall Tires, 6.70x15 nylon	−	−
1625	24 Gallon Fuel Tank	−	−

• A 283ci, 230hp engine, 3-speed manual transmission, vinyl interior trim, and a soft top were included in the base price.

• RPO 684 included special front and rear springs and shock absorbers, heavier front stabilizer bar, quick steering adaptor, metallic brakes, finned brake drums, fresh air ducting to rear brakes (early only) and front brake air deflectors. RPO 469C or RPO 579D, RPO 675, and manual transmission were required when RPO 684 was ordered.

• RPO 276 included wheels which were one-half inch wider than stock, and small passenger-car hubcaps in place of standard full wheel discs.

• LPO 1625 required the hardtop without soft top because the fuel tank occupied part of the folding top storage area.

• RPO 419 (Auxiliary Hardtop) quantity of 5,481 included 1,695 in place of soft tops at no cost.

• RPO 675 (Positraction) required manual transmission. The 4,170 quantity included 1,362 3.70:1 ratio, 2,523 4.11:1 ratio, and 285 4.56:1 ratio.

1959 Colors

EXTERIOR	QTY	SOFT TOP	WHEELS	INTERIOR
Tuxedo Black	1,594	Bk-W	Black	B-Bk-R
Classic Cream	223	Bk-W	Black	Bk
Frost Blue	1,024	B-Bk-W	Black	B-R
Crown Sapphire	888	Bk-Tq-W	Black	Tq
Roman Red	1,542	Bk-W	Black	Bk-R
Snowcrest White	3,354	B-Bk-Tq-W	Black	B-Bk-R-Tq
Inca Silver	957	Bk-W	Black	Bk-R

• Suggested interiors shown. Other combinations were possible.

• Interiors and exteriors were not coded to individual cars. In 1959, five Corvettes were painted a nonstandard color, color combination, or left in primer. An additional 83 were exported and their color combinations are not known.

• Numbers of interior colors sold in 1959 are as follows: 1,303 blue, 1,181 turquoise, 5,124 red, 2,062 black.

• The 2,931 quantity for code 440 two-tone paint (contrasting cove) was split 805 Roman Red/white, 535 Snowcrest White/silver, 496 Tuxedo Black/silver, 420 Crown Sapphire/white, 361 Frost Blue/white, 220 Inca Silver/white, 89 Classic Cream/white.

• Headliners for all soft tops were black. In 1959, soft top colors sold were 2,190 black, 4,092 white, 412 blue, and 217 turquoise.

•This was the only year for availability of a turquoise soft top color.

Abbreviations: B=Blue, Bk=Black, R=Red, Tq=Turquoise, W=White.

1960 Corvette
Production: 10,261 convertibles

1960 Numbers

Vehicle: 00867S100001 through 00867S110261

Suffix: CQ: 283ci, 230hp, mt   CU: 283ci, 270hp, mt
CR: 283ci, 250hp, mt   DG: 283ci, 230hp, at
CS: 283ci, 290hp, mt   DJ: 283ci, 245hp, at
CT: 283ci, 245hp, mt

• Suffix codes CY (283ci, 275hp) and CZ (283ci, 315hp) were assigned to aluminum-head versions of two fuel injected engines for 1960. Heads were produced, but not sold to retail customers.

Block: 3756519: All

Head: 3774692: All

Carburetor: Carter 2613S #3741089: 283ci, 270hp, fc
Carter 2614S #3741090: 283ci, 270hp, rc
Carter 2626S #3744002: 283ci, 245hp, fc
Carter 2627S #3744004: 283ci, 245hp, rc
Carter 2818S #3756676: 283ci, 230hp, fd
Carter 3059S #3779178: 283ci, 230hp, sd

Fuel Injection: Rochester 7017200: 283ci, 250hp
Rochester 7017250: 283ci, 290hp
Rochester 7017300: 283ci, 290hp
Rochester 7017310: 283ci, 250hp
Rochester 7017320: 283ci, 290hp

Distributor: 1110891: 283ci, 245hp, 270hp   1110915: 283ci, 250hp
1110914: 283ci, 290hp   1110946: 283ci, 230hp

Generator: 1102043: 283ci, 230hp, 245hp, 250hp, 270hp
1102173: 283ci, 290hp

Ending Vehicle: Oct 59: 101168   Feb 60: 104360   Jun 60: 109149
Nov 59: 101454   Mar 60: 105711   Jul 60: 109846
Dec 59: 102059   Apr 60: 107011   Aug 60: 110261
Jan 60: 103158   May 60: 108167

Abbreviations: at=automatic transmission, ci=cubic inch, fc=front carburetor, fd=first design, hp=horsepower, mt=manual transmission, rc=rear carburetor, sd=second design.

1960 Facts

• The 1960 exterior appearance continued the smooth contours of the previous years, and it was the last to feature taillights formed into the rounded rear fenders. It was also the last with heavy "teeth" in the grill.

• Aluminum radiators (with top tanks) appeared first in 1960 Corvettes, but use in 1960 was limited to 270hp and 290hp engines.

• Some literature lists 1960's two fuel-injected engines as 275hp and 315hp and fitted with aluminum heads. Problems associated with casting the high-silicon-content aluminum led to cancellation early in production. These aluminum heads were built and supplied to Briggs Cunningham through Chevrolet Engineering for the Cunningham LeMans effort, but it is believed none were delivered on cars sold to retail customers.

• All 1960 fuel injected engines required manual transmissions. Previously, automatic transmissions could be combined with the lower-horsepower fuel injected engines.

• The base 230hp engines had painted steel valve covers. All optional engines had seven-fin cast alloy valve covers. All covers had straight-across mounting holes and all attached with Phillips-head screws.

• The optional 1960 windshield washer reservoir mounted on the left side for all carbureted engines, and on the right side for all fuel injected engines. Right side mountings were protected by heat shields.

• The passenger-side storage bin and grab bar continued as in 1959, but 1960's anodized aluminum trim insert had a red bar above the "Corvette" letters, and a blue bar below. Corvette letters were black.

1960 Options

CODE	DESCRIPTION	QTY	RETAIL $
867	Base Corvette Convertible	10,261	$3,872.00
101	Heater	9,808	102.25
102	AM Radio, signal seeking	8,166	137.75
107	Parking Brake Alarm	4,051	5.40
108	Courtesy Light	6,774	6.50
109	Windshield Washers	7,205	16.15
121	Temperature Controlled Radiator Fan	2,711	21.55
261	Sunshades	5,276	10.80
276	Wheels, 15x5.5 (5)	246	0.00
290	Whitewall Tires, 6.70x15	9,104	31.55
313	Powerglide Automatic Transmission	1,766	199.10
419	Auxiliary Hardtop	5,147	236.75
426	Power Windows	544	59.20
440	Two-Tone Exterior Paint	3,312	16.15
469	283ci, 245hp Engine (2x4 carburetors)	1,211	150.65
469C	283ci, 270hp Engine (2x4 carburetors)	2,364	182.95
473	Power Operated Folding Top	512	139.90
579	283ci, 250hp Engine (fuel injection)	100	484.20
579D	283ci, 290hp Engine (fuel injection)	759	484.20
675	Positraction Rear Axle	5,231	43.05
685	4-Speed Manual Transmission	5,328	188.30
686	Metallic Brakes	920	26.90
687	Heavy Duty Brakes and Special Steering	119	333.60
1408	Blackwall Tires, 6.70x15 nylon	—	15.75
1625A	24 Gallon Fuel Tank	—	161.40

• A 283ci, 230hp engine, 3-speed manual transmission, vinyl interior trim, and a soft top were included in the base price.

• RPO 687 included special front and rear shocks, air scoops/deflectors for front brakes and air scoops for rear brakes, metallic brake facings, finned brake drums with cooling fans, and quick-steering adaptor. RPO 469C or RPO 579D, RPO 675, and manual transmission were required.

• RPO 276 included wheels which were one-half inch wider than stock, and small passenger-car hubcaps in place of standard full wheel discs.

• LPO 1625A (24 gallon fuel tank) required hardtop without soft top because the tank occupied part of the folding top storage area.

• The 5,147 RPO-419 (auxiliary hardtop) quantity included 1,641 in place of soft tops at no charge.

• RPO 675 (Positraction) required manual transmission. 5,231 quantity included 1,548 3.70:1 ratio, 3,226 4.11:1 ratio, and 457 4.56:1 ratio.

1960 Colors

EXTERIOR	QTY	SOFT TOP	WHEELS	INTERIOR
Tuxedo Black	1,268	B-Bk-W	Black	B-Bk-R-T
Tasco Turquoise	635	B-Bk-W	Turquoise	Bk-T
Horizon Blue	766	B-Bk-W	Blue	B-Bk-R
Honduras Maroon	1,202	Bk-W	Maroon	Bk
Roman Red	1,529	Bk-W	Red	Bk-R
Ermine White	3,717	B-Bk-W	White	B-Bk-R-T
Sateen Silver	989	B-Bk-W	Silver	B-Bk-R-T
Cascade Green	140	B-Bk-W	Green	Bk

• Suggested interiors shown. Other combinations were possible.

• Interiors and exteriors were not coded to individual cars.

• The number of interiors sold in 1960 were 3,231 black; 4,920 red; 1,078 turquoise; 1,032 blue.

• The 3,312 quantity for code 440 two-tone paint (contrasting cove) was split 779 Roman Red/white, 572 Honduras Maroon/white, 488 Ermine White/silver, 383 Tasco Turquoise/white, 386 Tuxedo Black/silver, 359 Horizon Blue/white, 280 Sateen Silver/white, 65 Cascade Green/white.

• Fifteen 1960 models were a nonstandard color, combination, or primer.

• Data suggests all soft top colors were available with all exteriors.

• Blue soft tops were a lighter shade than blue (or blue/gray) interiors.

• Cascade Green was metallic and different than 1956-57 Cascade Green.

• Headliners for all soft tops were black. In 1960, soft top colors sold were 4,931 black, 4,947 white, and 333 blue.

Abbreviations: B=Blue, Bk=Black, R=Red, T=Turquoise, W=White.

1961 Corvette

Production: 10,939 convertibles

1961 Numbers

Vehicle: 10867S100001 through 10867S110939

Suffix: CQ: 283ci, 230hp, mt   CU: 283ci, 270hp, mt
CR: 283ci, 275hp, mt   DG: 283ci, 230hp, at
CS: 283ci, 315hp, mt   DJ: 283ci, 245hp, at
CT: 283ci, 245hp, mt

Block: 3756519: All   3789935: All (lp)

Head: 3774692: 283ci, 230hp, 245hp, 270hp
3782461: 283ci, 275hp, 315hp

Carburetor: Carter 2613S #3741089: 283ci, 270hp, fc, fd
Carter 2614S #3741090: 283ci, 270hp, rc
Carter 2626S #3744002: 283ci, 245hp, fc, fd
Carter 2627S #3744004: 283ci, 245hp, rc
Carter 3059S #3779178: 283ci, 230hp
Carter 3181S #3785554: 283ci, 245hp, fc, sd
Carter 3182S #3785552: 283ci, 270hp, fc, sd

Fuel Injection: Rochester 7017200 & 7017310: 283ci, 275hp
Rochester 7017320: 283ci, 315hp

Distributor: 1110891: 283ci, 245hp, 270hp   1110915: 283ci, 275hp
1110914: 283ci, 315hp   1110946: 283ci, 230hp

Generator: 1102043: 283ci, 230hp, 245hp, 270hp, 275hp
1102173: 283ci, fd
1102268: 283ci, 315hp, sd

Ending Vehicle: Sep 60: 101052   Jan 61: 105203   May 61: 108960
Oct 60: 102301   Feb 61: 105966   Jun 61: 110160
Nov 60: 103355   Mar 61: 106889   Jul 61: 110939
Dec 60: 104306   Apr 61: 107804

Abbreviations: at=automatic transmission, ci=cubic inch, fc=front carburetor, fd=first design, hp=horsepower, lp=late production, mt=manual transmission, rc=rear carburetor, sd=second design.

1961 Facts

• Exterior styling was facelifted for 1961. It was the first Corvette without heavy "teeth" in the grill area. The forward headlight bezels were painted body color. The rear was completely restyled with four taillights, now a Corvette tradition, but a new look for 1961.

• Reduction of the transmission tunnel width increased interior space.

• A new crossflow aluminum radiator design was introduced in 1961 and this unit included a separate side-mount expansion tank. However, 1960 style radiators, both copper and aluminum, were used early in 1961 production because the new radiators were not available. Records indicated it was Chevrolet's intent to use copper radiators only for base engines while supplies lasted.

• Windshield washers, courtesy light, sun shades, temperature-controlled radiator fan, and parking brake warning light all became standard equipment in 1961 models.

• Windshield washer reservoirs mounted on the left side, except for fuel injected engines. For fuel injected engines, reservoirs were mounted on the right side and were protected by heat shields.

• The base 230hp engines had painted steel valve covers. All optional engines had seven-fin cast alloy valve covers. All covers had straight-across mounting holes and all attached with Phillips-head screws.

• Exhausts exited below the body on 1961s, a change from all previous Corvettes which exited through the rear body panel or rear bumper.

• Door sills were redesigned as one-piece, instead of two-piece as in 1960.

• The lower rear soft top bow for 1961 was aluminum, not steel as before.

• The grill for 1961 was finished in argent silver.

• Aluminum cases for 4-speed transmissions were introduced in 1961.

1961 Options

CODE	DESCRIPTION	QTY	RETAIL $
867	Base Corvette Convertible	10,939	$3,934.00
101	Heater	10,671	102.25
102	AM Radio, signal seeking	9,316	137.75
242	Positive Crankcase Ventilation	—	5.40
276	Wheels, 15x5.5 (5)	337	0.00
290	Whitewall Tires, 6.70x15	9,780	31.55
313	Powerglide Automatic Transmission	1,458	199.10
353	283ci, 275hp Engine (fuel injection)	118	484.20
354	283ci, 315hp Engine (fuel injection)	1,462	484.20
419	Auxiliary Hardtop	5,680	236.75
426	Power Windows	698	59.20
440	Two-Tone Exterior Paint	3,351	16.15
468	283ci, 270hp Engine (2x4 carburetor)	2,827	182.95
469	283ci, 245hp Engine (2x4 carburetor)	1,175	150.65
473	Power Operated Folding Top	442	161.40
675	Positraction Rear Axle	6,915	43.05
685	4-Speed Manual Transmission	7,013	188.30
686	Metallic Brakes	1,402	37.70
687	Heavy Duty Brakes and Special Steering	233	333.60
1408	Blackwall Tires, 6.70x15 nylon	—	15.75
1625	24 Gallon Fuel Tank	—	161.40

• A 283ci, 230hp engine, 3-speed manual transmission, vinyl interior trim, and a soft top were included in the base price.

•This was the last year for use of 283ci engines in Corvettes.

• RPO 687 included special front and rear shocks, air scoops/deflectors for front brakes and air scoops for rear brakes, metallic brake facings, finned brake drums with cooling fans, and quick-steering adaptor. RPO 354 or 468, and RPO 675 were required when RPO 687 was ordered.

• This was the last model with "wide" whitewall tires (optional). Original tires included U. S. Royal Air Ride, B. F. Goodrich Silvertown, and Firestone Deluxe Champion.

• RPO 276 included wheels one-half inch wider than stock, and small 1960 passenger car hubcaps in place of standard full discs.

• LPO 1625 (24 gallon fuel tank) required the hardtop without soft top because the tank occupied part of the folding top storage area.

• The 1,458 RPO 313 (automatic transmission) quantity was split 1,226 with 230hp engines, 232 with 245hp engines.

• The 5,680 RPO 419 (auxiliary hardtop) quantity included 2,285 in place of soft tops at no charge.

• RPO 675 (Positraction) required manual transmission.

1961 Colors

EXTERIOR	QTY	SOFT TOP	WHEELS	INTERIOR
Tuxedo Black	1,340	Bk-W	Black	B-Bk-F-R
Ermine White	3,178	Bk-W	White	B-Bk-F-R
Roman Red	1,794	Bk-W	Red	Bk-R
Sateen Silver	747	Bk-W	Silver	B-Bk-R
Jewel Blue	855	Bk-W	Blue	B-Bk
Fawn Beige	1,363	Bk-W	Beige	Bk-F-R
Honduras Maroon	1,645	Bk-W	Maroon	Bk-F

• Suggested interiors shown. Other combinations were possible.

• Interior and exterior colors were not coded to individual cars.

• Interior colors sold were: 4,459 Red; 3,487 Black; 1,662 Fawn; 1,331 Blue.

• Contrasting cove colors were last available in 1961.

• The 3,351 quantity for code 440 two-tone paint (contrasting cove) was split 954 Roman Red/white, 647 Honduras Maroon/white, 429 Tuxedo Black/silver, 419 Jewel Blue/white, 385 Ermine White/silver, 358 Fawn Beige/white, 159 Sateen Silver/white.

• In 1961, seventeen Corvettes were painted a nonstandard color, color combination, or left in primer.

• Jewel Blue exterior paint was exclusive to 1961.

• Headliners for all soft tops were black. In 1961, soft top colors sold were 3,052 black and 5,602 white.

Abbreviations: B=Blue, Bk=Black, F=Fawn, R=Red, W=White

1962 Corvette
Production: 14,531 convertibles

1962 Numbers

Vehicle: 20867S100001 through 20867S114531

Suffix: RC: 327ci, 250hp, mt   RF: 327ci, 360hp, mt
           RD: 327ci, 300hp, mt   SC: 327ci, 250hp, at
           RE: 327ci, 340hp, mt   SD: 327ci, 300hp, at

Block: 3782870: All

Head: 3774692: 327ci, 250hp, ep   3782461: 327ci, 300hp, 340hp, 360hp
        3795896: 327ci, 250hp   3884520: 327ci, 250hp, uu

Carburetor: Carter 3190S #3788245: 327ci, 250hp, at, ep
               Carter 3191S #3788246: 327ci, 250hp
               Carter 3310S #3819207: 327ci, 300hp, at
               Carter 3269S #3797699: 327ci, 300hp(mt), 340hp

Fuel Injection: Rochester 7017355 (ep)   Rochester 7017360

Distributor: 1110984: 327ci, 250hp,300hp   1110990: 327ci, 360hp,ep
             1110985: 327ci, 340hp   1111011: 327ci, 360hp

Generator: 1102174: 327ci, 250hp, 300hp
           1102268: 327ci, 340hp, 360hp

Ending Vehicle: Aug 61: 100443   Jan 62: 106234   Jun 62: 113459
                 Sep 61: 100827   Feb 62: 107585   Jul 62: 114520
                 Oct 61: 102065   Mar 62: 109116   Aug 62: 114531
                 Nov 61: 103465   Apr 62: 110519
                 Dec 61: 104766   May 62: 112035

Abbreviations: at=automatic transmission, ci=cubic inch, ep=early production, hp=horsepower, mt=manual transmission, uu=uncertain usage.

1962 Facts

• Engine displacement for 1962 increased from 283ci to 327ci. The base engine had 250hp. Dual-four barrel carburetors were not available.

• Styling resembled 1961, but there were visual differences. The side cove lip on 1962s was formed by the fiberglass body panels (which were unique to the year), not accented by bright trim as before. Because of this, 1962 Corvettes could not be ordered with coves painted to contrast body color. Also, this was the first model with rocker panel moldings.

• The simulated vent treatment in the cove area was changed in 1962 to a single louver, replacing the triple-spear vent.

• The conventional trunk design of the 1962 Corvette was the last for many years. Models to follow had no external rear storage access until 1982 when a special "collector edition" model featured a hatch window. But a trunk comparable to 1962's did not reappear until 1998's convertible.

• This was the last year for electric generators and solid rear axles. Also the last for exposed headlights and optional power soft tops until 2005.

• An aluminum case for the Powerglide automatic transmission was first introduced in 1962 Corvettes.

• This was the first year to have tires with narrow whitewalls (optional).

• Tachometers were distributor-driven in all 1962s. Previous use of tach-drive distributors in Corvette V8s was limited to fuel injected engines.

• The 250hp and 300hp engines had painted steel valve covers. The 340hp and 360hp engines had seven-fin cast alloy valve covers. All covers had straight-across mounting holes and all attached with Phillips-head screws.

• All 1962s were equipped with aluminum cross flow radiators and separate aluminum expansion tanks. Radiators were painted gloss black.

• Windshield washer reservoirs on 1962s mounted on the left side except for fuel injected engines. For fuel injected engines, reservoirs mounted on the right and were protected by heat shields.

• The grill for 1962 was finished in black anodized, gold anodized, or gold anodized painted black.

1962 Options

CODE	DESCRIPTION	QTY	RETAIL $
867	Base Corvette Convertible	14,531	$4,038.00
102	AM Radio, signal seeking	13,076	137.75
203	Rear Axle, 3.08:1 ratio	—	0.00
242	Positive Crankcase Ventilation	—	5.40
276	Wheels, 15x5.5 (5)	561	0.00
313	Powerglide Automatic Transmission	1,532	199.10
396	327ci, 340hp Engine	4,412	107.60
419	Auxiliary Hardtop	8,074	236.75
426	Power Windows	995	59.20
441	Direct Flow Exhaust System	2,934	0.00
473	Power Operated Folding Top	350	139.90
488	24 Gallon Fuel Tank	65	118.40
582	327ci, 360hp Engine (fuel injection)	1,918	484.20
583	327ci, 300hp Engine	3,294	53.80
675	Positraction Rear Axle	14,232	43.05
685	4-Speed Manual Transmission	11,318	188.30
686	Metallic Brakes	2,799	37.70
687	Heavy Duty Brakes and Special Steering	246	333.60
1832	Whitewall Tires, 6.70x15	—	31.55
1833	Blackwall Tires, 6.70x15 nylon	—	15.70

• A 327ci, 250hp engine, 3-speed manual transmission, vinyl interior trim, and a soft top were included in the base price.

• Heaters became standard in 1962 for the first time, but could be factory-deleted with RPO 610. The 610 code specified "export," but other heater-deletes were built for racing.

• RPO 687 included special front and rear shocks, air scoops/deflectors for front brakes and air scoops for rear brakes, metallic brake facings, finned brake drums with cooling fans, and quick-steering adaptor. RPOs 582 and 675 were required when RPO 687 was ordered.

• Original tires included U. S. Royal Air Ride, B. F. Goodrich Silvertown, Firestone Deluxe Champion, Goodyear Custom Super Cushion, and General Jet Air. Optional whitewalls were new thin style, varying in width from slightly under to slightly over an inch.

• RPO 242 (pcv) was specified in order guides to be for California use.

• RPO 488 (24-gallon fuel tank) required hardtop without soft top because the tank occupied part of the folding top storage area.

• RPO 276 included wheels one-half inch wider than stock, and small 1960 passenger car hubcaps in place of standard full discs.

• Base 3-speed manual transmissions were split 1,619 with 250hp/300hp engines, 62 with 340hp/360hp engines.

• The 1,532 RPO 313 quantity was split 1,067 with 250hp engines, 465 with 300hp engines.

• The 8,074 RPO 419 quantity included 3,179 in place of soft tops.

• The 11,318 RPO 685 quantity was split 5,050 with 250hp/300hp engines, 6,268 with 340hp/360hp engines.

1962 Colors

EXTERIOR	QTY	SOFT TOP	WHEELS	INTERIOR
Tuxedo Black	—	Bk-W	Bk	Bk-F-R
Fawn Beige	1,851	Bk-W	Bk-Fb	F-R
Roman Red	—	Bk-W	Bk-R	Bk-F-R
Ermine White	—	Bk-W	Bk-W	Bk-F-R
Almond Beige	820	Bk-W	Bk-Ab	F-R
Sateen Silver	—	Bk-W	Bk-Si	Bk-R
Honduras Maroon	—	Bk-W	Bk-M	Bk-F

• Suggested interiors shown. Other combinations were possible.

• Generally, 1962s with whitewall tires had black wheels. Wheels combined with blackwall tires or RPO 276 were painted body color.

• Headliners for soft tops were black. In 1962, soft top colors sold were 4,727 black, and 6,625 white.

• Interior and exterior colors were not coded to individual cars. Other exterior colors, including primer, were sold. For example, Cadillac Royal Heather Amethyst (purple) is documented for a small number of 1962 Corvettes built for Omaha, Nebraska's Tangier Shrine Corvette Patrol.

Abbreviations: Ab=Almond Beige, Bk=Black, F=Fawn, Fb=Fawn Beige, M=Maroon, R=Red, Si=Silver, W=White.

1963 Corvette

Production: 10,594 coupe, 10,919 convertible, 21,513 total.

1963 Numbers

Vehicle: 30837S100001 through 30837S121513
  • For convertibles, fourth digit is a 6.

Suffix: RC: 327ci, 250hp, mt   RF: 327ci, 360hp, mt
RD: 327ci, 300hp, mt   SC: 327ci, 250hp, at
RE: 327ci, 340hp, mt   SD: 327ci, 300hp, at

Block: 3782870: All

Head: 3795896: 327ci, 250hp
3782461: 327ci, 300hp, 340hp, 360hp

Carburetor: Carter 3460S #3826006: 327ci, 300hp, at
Carter 3461S #3826004: 327ci, 300hp, 340hp, mt
Carter 3500S #3826005: 327ci, 250hp, at
Carter 3501S #3826003: 327ci, 250hp, mt

Fuel Injection: Rochester 7017375

Distributor: 1111022: 327ci, 360hp
1111024: 327ci, 250hp, 300hp, 340hp

Alternator: 1100628: All without ac
1100633: All with ac

Ending Vehicle: Sep 62: 100675   Jan 63: 107976   May 63: 116409
Oct 62: 102756   Feb 63: 109814   Jun 63: 118524
Nov 62: 104047   Mar 63: 111833   Jul 63: 120990
Dec 62: 105972   Apr 63: 114128   Aug 63: 121513

Abbreviations: ac=air conditioning, at=automatic transmission, ci=cubic inch, hp=horsepower, mt=manual transmission.

1963 Facts

• For 1963, the Corvette's body and chassis were completely redesigned. For the first time, a coupe body was available. A center wind split on the coupe roof divided the rear glass creating a "split window," unique to this year. The new chassis featured an independent rear suspension with a single transverse leaf spring. Engines carried over from 1962.

• Knock-off aluminum wheels were introduced as a 1963 option, but actual availability is questionable. Porosity of the aluminum and rim seal difficulty in early wheels caused tubeless tires to leak. Delivery of a 1963 with knock-off wheels to a retail customer has not been confirmed, but wheels were sold over-the-counter. Two bar (early) and three-bar spinner styles were available. Finish between the fins was natural.

• The 1963 exterior doors had raised pads for the door handles. Also, coupes had stainless steel trim forward of the vent window.

• Instrument faces had black faces with deep, bright-finish center recesses.

• Inside door release knobs and shift knobs were black plastic.

• Powerglide automatic transmissions had staggered shift gates.

• Early 1963 radios were AM signal-seeking. Later were AM-FM.

• All 1963 Corvettes had built-in adjusting mechanisms for the bottom seat cushions. Early 1963s had under-seat depressions, possibly for tool storage. This feature was removed about midway during the model year.

• Most 1963 Corvettes had fiberglass headlight buckets. Late 1963s and all 1964-67 models had metal buckets.

• Early 1963 models used roller-type catches for the gas filler doors. Later production used nylon slide catches.

• The glove box door in the 1963 Corvette was fiberglass and its face was covered with clear plastic. In early 1963s, the dash surface around the radio and speaker bezel was painted instead of vinyl-covered.

• 1963 hoods had simulated air vent panels in two forward recesses.

• 4-speed manual transmissions changed from Borg-Warner manufacture to Muncie during the 1963 model year.

• The outside rearview mirror was revised to a taller design with a smaller base at about the midpoint of 1963 production.

1963 Options

RPO #	DESCRIPTION	QTY	RETAIL $
837	Base Corvette Sport Coupe	10,594	$4,252.00
867	Base Corvette Convertible	10,919	4,037.00
898	Genuine Leather Seats	1,114	80.70
941	Sebring Silver Exterior Paint	3,516	80.70
A01	Soft Ray Tinted Glass, all windows	629	16.15
A02	Soft Ray Tinted Glass, windshield	470	10.80
A31	Power Windows	3,742	59.20
C07	Auxiliary Hardtop (for convertible)	5,739	236.75
C48	Heater and Defroster Deletion (credit)	124	-100.00
C60	Air Conditioning	278	421.80
G81	Positraction Rear Axle, all ratios	17,554	43.05
G91	Special Highway Axle, 3.08:1 ratio	211	2.20
J50	Power Brakes	3,336	43.05
J65	Sintered Metallic Brakes	5,310	37.70
L75	327ci, 300hp Engine	8,033	53.80
L76	327ci, 340hp Engine	6,978	107.60
L84	327ci, 360hp Engine (fuel injection)	2,610	430.40
M20	4-Speed Manual Transmission	17,973	188.30
M35	Powerglide Automatic Transmission	2,621	199.10
N03	36 Gallon Fuel Tank (for coupe)	63	202.30
N11	Off Road Exhaust System	—	37.70
N34	Woodgrained Plastic Steering Wheel	130	16.15
N40	Power Steering	3,063	75.35
P48	Cast Aluminum Knock-Off Wheels (5)	—	322.80
P91	Blackwall Tires, 6.70x15, (nylon cord)	412	15.70
P92	Whitewall Tires, 670x15 (rayon cord)	19,383	31.55
T86	Back-up Lamps	318	10.80
U65	Signal Seeking AM Radio	11,368	137.75
U69	AM-FM Radio	9,178	174.35
Z06	Special Performance Equipment	199	1,818.45

• A 327ci, 250hp engine, 3-speed manual transmission, vinyl interior trim, and a soft top (convertible) were included in the base price.

• Z06 was initially coupe-only. Later, Z06 cost $1,293.95 and excluded knock-off wheels and 36 gallon tank, and was available with convertibles.

• The 5,739 C07 quantity included 1,099 in place of soft tops (no cost).

• The 17,554 G81 quantity included 2,259 3.08:1 ratio, 6,855 3.36:1 ratio, 613 3.55:1 ratio, 2,570 3.70:1 ratio, 4,506 4.11:1 ratio, and 751 4.56:1 ratio.

• The 2,621 M35 quantity was split 1,116 with 250hp, 1,505 with 300hp.

• Delivery of a 1963 to a retail customer with RPO P48 cast aluminum wheels has not been verified.

• RPO U69 radios were phased in around March 1963, but these and RPO U65 radios were available simultaneously as supplies permitted.

1963 Colors

CODE	EXTERIOR	QTY	SOFT TOP	WHEELS	INTERIORS
900	Tuxedo Black	—	Bk-W-Bg	Bk	Bk-R-S
912	Silver Blue	—	Bk-W-Bg	Bk-Si	Bk-Db
916	Daytona Blue	3,475	Bk-W-Bg	Bk-Db	Db-R-S
923	Riverside Red	4,612	Bk-W-Bg	Bk-R	Bk-R-S
932	Saddle Tan	—	Bk-W-Bg	Bk-S	Bk-R-S
936	Ermine White	—	Bk-W-Bg	Bk-W	Bk-Db-R-S
941	Sebring Silver	3,516	Bk-W-Bg	Bk-Si	Bk-Db-R-S

• Suggested interiors shown. Other combinations were possible.

• When whitewall tires were ordered, the standard wheels were painted black. With blackwalls, wheels were painted body color (white exteriors may have had black wheels regardless of tire type).

• Soft top color quantities were 3,648 black, 5,728 white, 444 beige.

Interior Codes: Std/Blk=Bk, 490A/J/S/XE/XG=Db/V-cpe, 490B/K/T/XF/XH=Db/V-con, 490C/L/Q/XA/XC=R/V-cpe, 490D/M/R/XB/XD=R/V-con, 490E/N/U/XJ/XL=S/V-cpe, 490F/P/V/XK/XM=S/V-con, 898A/E/Q/G/S=S/L-cpe, 898B/F/R/H/T=S/L-con.

• With the exception of "std" or "blk" for black vinyl, codes had three numbers followed by a one or two alpha-character suffix. This was the Corvette's first year for coding of trim tags and inconsistencies exist.

Abbreviations: Bg=Beige, Bk=Black, con=convertible, cpe=coupe, Db=Dark Blue, L=Leather, R=Red, S=Saddle, Si=Silver, V=Vinyl, W=White.

1964 Corvette

Production: 8,304 coupe, 13,925 convertible, 22,229 total.

1964 Numbers

Vehicle: 40837S100001 through 40837S122229
  • For convertibles, fourth digit is a 6.

Suffix: RC: 327ci, 250hp, mt   RT: 327ci, 365hp, mt, ig
  RD: 327ci, 300hp, mt   RU: 327ci, 365hp, mt, ac, ig
  RE: 327ci, 365hp, mt   RX: 327ci, 375hp, mt, ig
  RF: 327ci, 375hp, mt   SC: 327ci, 250hp, at
  RP: 327ci, 250hp, mt, ac   SD: 327ci, 300hp, at
  RQ: 327ci, 300hp, mt, ac   SK: 327ci, 250hp, at, ac
  RR: 327ci, 365hp, mt, ac   SL: 327ci, 300hp, at, ac

Block: 3782870: All

Head: 3782461: 327ci, 300hp, 365hp, 375hp
  3795896: 327ci, 250hp

Carburetor: Carter 3696S #3846246: 327ci, 250hp, at
  Carter 3697S #3846247: 327ci, 250hp, mt
  Carter 3720S/SA/SB #3851762: 327ci, 300hp, at
  Carter 3721S/SA/SB #3851761: 327ci, 300hp, mt
  Holley R2818A #3849804: 327ci, 365hp

Fuel Injection: Rochester 7017375, ep   Rochester 7017380, lp
  Rochester 7017375R, lu

Distributor: 1111024: 327ci, 250hp,300hp   1111064: 327ci, 375hp, ig
  1111060: 327ci, 365hp, ig   1111069: 327ci, 365hp, lp
  1111062: 327ci, 365hp, ep   1111070: 327ci, 375hp, lp
  1111063: 327ci, 375hp, ep

Alternator: 1100628: 327ci, ep, lu
  1100665: 327ci, 250hp, 300hp, 365hp, ac
  1100668: 327ci, 250hp, 300hp, 365hp, 375hp
  1100669: 327ci, 365hp, 375hp, ig
  1100684: 327ci, 365hp, ac, ig

Ending Vehicle: Sep 63: 101741   Jan 64: 110297   May 64: 118805
  Oct 63: 104045   Feb 64: 112322   Jun 64: 120920
  Nov 63: 106063   Mar 64: 114570   Jul 64: 122229
  Dec 63: 108091   Apr 64: 116865

Abbreviations: ac=air conditioning, at=automatic transmission, ci=cubic inch, ep=early production, hp=horsepower, ig=transistor ignition, lp=late production, lu=limited use, mt=manual transmission.

1964 Facts

• The split window in coupes was eliminated, so rear glass was one piece. Also, the simulated vent panels in the hood were removed, but the recesses remained making 1964's hood unique.

• A three-speed fan was added to improve ventilation in coupes. The fan exhausted interior air through functional vents added to the driver side roof panel (passenger side vents weren't functional). The control switch was mounted under the driver's side of the instrument panel.

• Seats in 1964s looked similar to 1963, but 1964 backs were thicker and less tapered at the top. Also, 1964 seats didn't have adjusting mechanisms for the bottom cushions as in 1963.

• Delivery of the knock-off wheel option during 1964 was certain. Only three-bar spinners were offered and the finish between fins was natural.

• Center recess areas of 1964 instruments were finished in black.

• Steering wheel rims in all 1964 Corvettes were walnut-grained plastic.

• The exterior door surface of 1964 models had a raised "pad" for the door handle to mount (except for late 1964s).

• Inside door release knobs were chromed.

• Powerglide automatic transmissions had staggered shift gates

• Starting in 1964 (through 1967), some Corvette bodies were supplied by Dow-Smith, Ionia, Michigan, a division of A. O. Smith Company.

1964 Options

RPO #	DESCRIPTION	QTY	RETAIL $
837	Base Corvette Sport Coupe	8,304	$4,252.00
867	Base Corvette Convertible	13,925	4,037.00
—	Genuine Leather Seats	1,334	80.70
A01	Soft Ray Tinted Glass, all windows	6,031	16.15
A02	Soft Ray Tinted Glass, windshield	6,387	10.80
A31	Power Windows	3,706	59.20
C07	Auxiliary Hardtop (for convertible)	7,023	236.75
C48	Heater and Defroster Deletion (credit)	60	-100.00
C60	Air Conditioning	1,988	421.80
F40	Special Front and Rear Suspension	82	37.70
G81	Positraction Rear Axle, all ratios	18,279	43.05
G91	Special Highway Axle, 3.08:1 ratio	2,310	2.20
J50	Power Brakes	2,270	43.05
J56	Special Sintered Metallic Brake Package	29	629.50
J65	Sintered Metallic Brakes, power	4,780	53.80
K66	Transistor Ignition System	552	75.35
L75	327ci, 300hp Engine	10,471	53.80
L76	327ci, 365hp Engine	7,171	107.60
L84	327ci, 375hp Engine (fuel injection)	1,325	538.00
M20	4-Speed Manual Transmission	19,034	188.30
M35	Powerglide Automatic Transmission	2,480	199.10
N03	36 Gallon Fuel Tank (for coupe)	38	202.30
N11	Off Road Exhaust System	1,953	37.70
N40	Power Steering	3,126	75.35
P48	Cast Aluminum Knock-Off Wheels (5)	806	322.80
P91	Blackwall Tires, 6.70x15 (nylon cord)	372	15.70
P92	Whitewall Tires, 6.70x15 (rayon cord)	19,977	31.85
T86	Back-up Lamps	11,085	10.80
U69	AM-FM Radio	20,934	176.50

• A 327ci, 250hp engine, 3-speed manual transmission, vinyl interior trim, and a soft top (convertible) were included in the base price.

• The 7,023 C07 quantity included 1,220 in place of soft tops (no cost).

• The 1,988 C60 quantity was split 1,069 coupe, 919 convertible.

• The 18,279 G81 quantity included 2,475 3.08:1 ratio, 8,338 3.36:1 ratio, 486 3.55:1 ratio, 2,360 3.70:1 ratio, 3,979 4.11:1 ratio, and 641 4.56:1 ratio.

• RPO J56 required fuel injected engine, 4-speed, and Positraction.

• The 19,034 M20 quantity was split 10,538 with wide ratio 250hp and 300hp engines; 8,496 with close ratio 365hp and 375hp engines

• The 2,480 M35 quantity was split 904 with 250hp, 1,576 with 300hp.

1964 Colors

CODE	EXTERIOR	QTY	SOFT TOP	WHEELS	INTERIORS
900	Tuxedo Black	1,897	Bk-W-Bg	Black	Bk-R-Si-W
912	Silver Blue	3,121	Bk-W-Bg	Black	Bk-B-W
916	Daytona Blue	3,454	Bk-W-Bg	Black	B-Si-W
923	Riverside Red	5,274	Bk-W-Bg	Black	Bk-R-W
932	Saddle Tan	1,765	Bk-W-Bg	Black	S-W
936	Ermine White	3,909	Bk-W-Bg	Black	Bk-B-R-S-Si-W
940	Satin Silver	2,785	Bk-W-Bg	Black	Bk-B-R-Si-W

• Suggested interiors shown. Other combinations were possible.

• In 1964, 24 Corvettes had non-standard paint, or primer.

• Soft top color quantities were 4,721 black, 4,843 white, 591 beige.

Interior Codes: Std=Bk/V, 898A=Bk/L, 490AA/AB/G/H=R/V, 898FA/ EA/L/M=R/L, 490BA/BB/J/K=B/V, 898JA/KA/N/P=B/L, 490CA/ CB/L/M=S/V, 898CA/DA/G/H=S/L, 491AA/AE=Si+Bk/V, 899AA/ AE=Si+Bk/L, 491BA/BE/M/N=Si+B/V, 899BA/BE/M/N=Si+B/L, 491CA/CB/CE=W+Bk/V, 899CA/CB/CE=W+Bk/L, 491GA/GE/R/ S=W+B/V, 899GA/GE/R/S=W+B/L, 491DA/DE/P/Q=W+R/V, 899DA/DE/P/Q=W+R/L, 491HA/HE/T/U=W+S/V, 899HA/HE/T/ U=W+S/L.

• With the exception of "std" for black vinyl, 1964 interior codes consisted of three numbers followed by a one or two alpha-character suffix. In some cases, the suffix differentiated coupe from convertible and/or A. O. Smith body build from St. Louis in-house body build.

Abbreviations: B=Blue, Bg=Beige, Bk=Black, L=Leather, R=Red, S=Saddle, Si=Silver, V=Vinyl, W=White.

1965 Corvette

Production: 8,186 coupe, 15,378 convertible, 23,564 total.

1965 Numbers

Vehicle: 194375S100001 through 194375S123564
  • For convertibles, fourth digit is a 6.

Suffix:

HE: 327ci, 250hp, mt	HO: 327ci, 250hp, at
HF: 327ci, 300hp, mt	HP: 327ci, 300hp, at
HG: 327ci, 375hp, mt	HQ: 327ci, 250hp, at, ac
HH: 327ci, 365hp, mt	HR: 327ci, 300hp, at, ac
HI: 327ci, 250hp, mt, ac	HT: 327ci, 350hp, mt
HJ: 327ci, 300hp, mt, ac	HU: 327ci, 350hp, mt, ac
HK: 327ci, 365hp, mt, ac	HV: 327ci, 350hp, mt, ig
HL: 327ci, 365hp, mt, ig	HW: 327ci, 350hp, mt, ac, ig
HM: 327ci, 365hp, mt, ac, ig	IF: 396ci, 425hp, mt, ig
HN: 327ci, 375ph, mt, ig	

Block: 3782870: 327ci   3855962: 396ci   3858180: 327ci (lu)

Head: 3782461: 327ci   3856208: 396ci

Carburetor: Carter 3696S #3846246: 327ci, 250hp, at
  Carter 3697S #3846247: 327ci, 250hp, mt
  Carter 3720SB #3851762: 327ci, 300hp, at
  Carter 3721SB #3851761: 327ci, 300hp, mt
  Holley R2818A #3849804: 327ci, 350hp, 365hp
  Holley R3124A #3868826: 396ci, 425hp

Fuel Injection: Rochester 7017380

Distributor: 1111060: 327ci, 365hp, ig   1111076: 327ci, 250hp, 300hp
  1111064: 327ci, 375hp, ig   1111087: 327ci, 350hp
  1111069: 327ci, 365hp   1111088: 327ci, 350hp, ig
  1111070: 327ci, 375hp   1111093: 396ci, 425hp, ig

Alternator: 1100693: 327ci   1100696: 327ci, 396ci, ig
  1100694: 327ci, ac   1100697: 327ci, 350hp, 365hp, ac, ig

Ending Vehicle:

Aug 64: 100227	Jan 65: 108442	May 65: 118753
Sep 64: 101425	Feb 65: 111059	Jun 65: 121216
Nov 64: 103347	Mar 65: 113936	Jul 65: 123562
Dec 64: 105754	Apr 65: 116516	Aug 65: 123564

Abbreviations: ac=air conditioning, at=automatic transmission, ci=cubic inch, hp=horsepower, ig=transistor ignition, lu=limited use, mt=manual transmission.

1965 Facts

• Hood depressions common to the 1963-64 models were removed. Horizontal grill bars changed to black, but the outer grill trim remained bright, making 1965's grill unique. The side front fender louvers were changed to three functional, vertical slots.

• Chevrolet built two 1965s after production officially ended July 31. Because of this, 1965 production has been incorrectly reported as 23,562 instead of 23,564. Counts on the opposing page are based on 23,562.

• This was the last year for fuel injection in Corvettes until throttle-body injection in 1982. In mid-March 1965, the big-block engine option was introduced. It came in one version, 396ci with 425hp, and included a new hood design with an exaggerated central bulge.

• Four-wheel disc brakes were introduced as standard equipment for 1965 Corvettes. The system included four-piston caliper assemblies at each wheel. Although disc brakes were included in 1965's price, drum brakes could be ordered as a credit ($64.50) option while supplies lasted.

• Instruments were restyled with black, flat faces. The area around the radio and speaker bezel was painted, rather than vinyl-covered.

• Optional knock-off wheels were painted dark grey between the fins.

• A power antenna, Corvette's first, was standard equipment.

• Seats were redesigned. Seating surfaces were slightly larger and more supportive, and backs were encased in hard plastic backing shells. Inner door panels were also redesigned with integrated arm rests.

1965 Options

RPO #	DESCRIPTION	QTY	RETAIL $
19437	Base Corvette Sport Coupe	8,186	$4,321.00
19467	Base Corvette Convertible	15,378	4,106.00
—	Genuine Leather Seats	2,128	80.70
A01	Soft Ray Tinted Glass, all windows	8,752	16.15
A02	Soft Ray Tinted Glass, windshield	7,624	10.80
A31	Power Windows	3,809	59.20
C07	Auxiliary Hardtop (for convertible)	7,787	236.75
C48	Heater and Defroster Deletion (credit)	39	-100.00
C60	Air Conditioning	2,423	421.80
F40	Special Front and Rear Suspension	975	37.70
G81	Positraction Rear Axle, all ratios	19,965	43.05
G91	Special Highway Axle, 3.08:1 ratio	1,886	2.20
J50	Power Brakes	4,044	43.05
J61	Drum Brakes (substitution credit)	316	-64.50
K66	Transistor Ignition System	3,686	75.35
L75	327ci, 300hp Engine	8,358	53.80
L76	327ci, 365hp Engine	5,011	129.15
L78	396ci, 425hp Engine	2,157	292.70
L79	327ci, 350hp Engine	4,716	107.60
L84	327ci, 375hp Engine (fuel injection)	771	538.00
M20	4-Speed Manual Transmission	21,107	188.30
M35	Powerglide Automatic Transmission	2,021	199.10
N03	36 Gallon Fuel Tank (for coupe)	41	202.30
N11	Off Road Exhaust System	2,468	37.70
N14	Side Mount Exhaust System	759	134.50
N32	Teakwood Steering Wheel	2,259	48.45
N36	Telescopic Steering Column	3,917	43.05
N40	Power Steering	3,236	96.85
P48	Cast Aluminum Knock-Off Wheels (5)	1,116	322.80
P91	Blackwall Tires, 7.75x15 (nylon cord)	168	15.70
P92	Whitewall Tires, 7.75x15 (rayon cord)	19,300	31.85
T01	Goldwall Tires, 7.75x15 (nylon cord)	989	50.05
U69	AM-FM Radio	22,113	203.40
Z01	Comfort and Convenience Group	15,397	16.15

• A 327ci, 250hp engine, 3-speed manual transmission, vinyl interior trim, and soft top (convertible) were included in the base price.

• The 7,787 C07 quantity included 1,277 in place of soft tops (no cost).

• The 2,423 C60 (air conditioning) quantity was split 1,551 coupe, 872 convertible. It was not available with RPOs L78 or L84.

• The 19,965 G81 quantity included 2,140 3.08:1 ratio, 7,066 3.36:1 ratio, 1,685 3.55:1 ratio, 3,991 3.70:1 ratio, 4,294 4.11:1 ratio, and 789 4.56:1 ratio.

• RPO K66 (Transistor Ignition) was required purchase with RPO L78.

• The 2,021 M35 quantity was split 663 with 250hp, 1,358 with 300hp.

• The P91 blackwall tire option was a nylon replacement for the standard rayon blackwall. It was cancelled shortly after 1965 production started.

1965 Colors

CODE	EXTERIOR	QTY	SOFT TOP	WHEELS	INTERIORS
AA	Tuxedo Black	1,191	Bk-W-Bg	Black	Bk-B-G-M-R-S-Si-W
CC	Ermine White	2,216	Bk-W-Bg	Black	Bk-B-G-M-R-S-Si-W
FF	Nassau Blue	6,022	Bk-W-Bg	Black	Bk-B-W
GG	Glen Green	3,782	Bk-W-Bg	Black	Bk-G-S-W
MM	Milano Maroon	2,831	Bk-W-Bg	Black	Bk-R-M-S-W
QQ	Silver Pearl	2,552	Bk-W-Bg	Black	Bk-R-Si
UU	Rally Red	3,688	Bk-W-Bg	Black	Bk-R-W
XX	Goldwood Yellow	1,275	Bk-W-Bg	Black	Bk-W

• Suggested interiors shown. Other combinations were possible.

• Codes for St.Louis-built bodies had a 900 prefix. A.O.Smith-built bodies had no prefix and dash (-) between code letters.

• The first 1965 was painted 1964 Satin Silver and coded ZZ (Cadillac's code for the same color). Five 1965s had nonstandard paint, or primer.

• Soft top color quantities were 5,565 black, 7,983 white, 551 beige.

Interior Codes: Std=Bk/V, 402=Bk/L, 407=R/V, 408=R/L, 414=B/V, 415=B/L, 420=S/V, 421=S/L, 426=Si/V, 427=Si/L, 430=G/V, 431=G/L, 435=M/V, 436=M/L, 437=W+Bk/V, 438=W+Bk/L, 443=W+R/V, 444=W+R/L, 450=W+B/V, 451=W+B/L.

Abbreviations: B=Blue, Bg=Beige, Bk=Black, G=Green, L=Leather, M=Maroon, R=Red, S=Saddle, Si=Silver, V=Vinyl, W=White.

1966 Corvette

Production: 9,958 coupe, 17,762 convertible, 27,720 total.

1966 Numbers

Vehicle: 194376S100001 through 194376S127720
  • For convertibles, fourth digit is a 6.

Suffix: HD: 327ci, 350hp, mt, ar
  HE: 327ci, 300hp, mt
  HH: 327ci, 300hp, mt, ar
  HO: 327ci, 300hp, at
  HP: 327ci, 350hp, mt, ps, ac
  HR: 327ci, 300hp, at, ar
  HT: 327ci, 350hp, mt

  IK: 427ci, 425hp, mt
  IL: 427ci, 390hp, mt
  IM: 427ci, 390hp, mt, ar
  IP: 427ci, 425hp, mt
  IQ: 427ci, 390hp, at
  IR: 427ci, 390hp, at, ar
  KH: 327ci, 350hp, mt, ar, ac, ps

Block: 3858174: 327ci   3892657: 327ci, lp (uu)
  3869942: 427ci   3855961: 427ci, ep (uu)

Head: 3782461: 327ci, 300hp, 350hp   3873858: 427ci, 425hp
  3872702: 427ci, 390hp

Carburetor: Holley R3247A #3886101: 427ci, 425hp
  Holley R3367A #3884505: 327ci, 300hp, 350hp
  Holley R3370A #3882835: 427ci, 390hp
  Holley R3605A #3890499: 327ci, 300hp, 350hp, ar
  Holley R3606A #3892341: 427ci, 390hp, ar

Distributor: 1111093: 427ci, 425hp, ig   1111153: 327ci, 300hp
  1111141: 427ci, 390hp   1111156: 327ci, 350hp
  1111142: 427ci, 390hp, ig   1111157: 327ci, 350hp, ig

Alternator: 1100693: 327ci, 427ci   1100696: 327ci, 427ci, ig
  1100694: 327ci, 427ci, ac   1100750: 327ci, 427ci, ac, ig, uu

Ending Vehicle: Sep 65: 102031   Jan 66: 112587   May 66: 123016
  Oct 65: 104384   Feb 66: 115283   Jun 66: 125469
  Nov 65: 107186   Mar 66: 118091   Jul 66: 127720
  Dec 65: 109892   Apr 66: 120664

Abbreviations: ac=air conditioning, ar=air injection reactor, at=automatic transmission, ci=cubic inch, ep=early production, hp=horsepower, ig=transistor ignition, lp=late production, mt=manual transmission, ps=power steering, uu=uncertain usage.

1966 Facts

• The 1966 Corvette's styling was similar to the previous model, but there were subtle differences, including the addition of the Corvette script (elongated vertical style) to the hood, and a plated, square-mesh, cast grill. Also, roof vents which had been both functional and non-functional in previous mid-year Corvette coupes, were deleted completely.

• Seats were similar to 1965, except 1966 seats had additional pleats in the upper and lower sections for better weight distribution at the seams.

• The knock-off wheel option continued in 1966, but with a dull-finish center cone instead of bright. The area between the fins was painted dark grey.

• The 427 cubic-inch engines were introduced in the 1966 model. Corvettes with these engines received the special bubble hoods first seen on 1965's with 396ci engines. The 1966 high performance, solid-lifter 427 was initially listed and labeled at 450hp but, for uncertain reasons, the rating was reduced to 425hp shortly after introduction. Similarly, the 390hp engine was initially rated at 400hp. These were administrative decisions not related to any changes in the engines.

• Backup lights became standard equipment in the 1966 model. They were incorporated into the existing rear inboard taillight housings.

• The fiberboard headliners of coupes and convertible hardtops (except early) were replaced with vinyl-covered foam. Interior door pulls were bright metal. Headrests were optional for the first time.

• Holley carburetors, used with some engine applications previously, were standard with all 1966 engines.

• Exhaust bezels were plated castings, not stainless as earlier mid years.

1966 Options

RPO #	DESCRIPTION	QTY	RETAIL$
19437	Base Corvette Sport Coupe	9,958	$4,295.00
19467	Base Corvette Convertible	17,762	4,084.00
—	Genuine Leather Seats	2,002	79.00
A01	Soft Ray Tinted Glass, all windows	11,859	15.80
A02	Soft Ray Tinted Glass, windshield	9,270	10.55
A31	Power Windows	4,562	57.95
A82	Headrests	1,033	42.15
A85	Shoulder Belts	37	26.35
C07	Auxiliary Hardtop (for convertible)	8,463	231.75
C48	Heater and Defroster Deletion (credit)	54	-97.85
C60	Air Conditioning	3,520	412.90
F41	Special Front and Rear Suspension	2,705	36.90
G81	Positraction Rear Axle, all ratios	24,056	42.15
J50	Power Brakes	5,464	42.15
J56	Special Heavy Duty Brakes	382	342.30
K19	Air Injection Reactor	2,380	44.75
K66	Transistor Ignition System	7,146	73.75
L36	427ci, 390hp Engine	5,116	181.20
L72	427ci, 425hp Engine	5,258	312.85
L79	327ci, 350hp Engine	7,591	105.35
M20	4-Speed Manual Transmission	10,837	184.35
M21	4-Speed Man Trans, close ratio	13,903	184.35
M22	4-Speed Man Trans, close ratio, heavy duty	15	237.00
M35	Powerglide Automatic Transmission	2,401	194.85
N03	36 Gallon Fuel Tank (for coupe)	66	198.05
N11	Off Road Exhaust System	2,795	36.90
N14	Side Mount Exhaust System	3,617	131.65
N32	Teakwood Steering Wheel	3,941	47.40
N36	Telescopic Steering Column	3,670	42.15
N40	Power Steering	5,611	94.80
P48	Cast Aluminum Knock-Off Wheels (5)	1,194	316.00
P92	Whitewall Tires, 7.75x15, (rayon cord)	17,969	31.30
T01	Goldwall Tires, 7.75x15 (nylon cord)	5,557	46.55
U69	AM-FM Radio	26,363	199.10
V74	Traffic Hazard Lamp Switch	5,764	11.60

• A 327ci, 300hp engine, 3-speed manual transmission, vinyl interior trim, and soft top (convertible) were included in the base price.

• The 8,463 C07 quantity included 1,303 in place of soft tops (no cost).

• The 3,520 C60 quantity was split 2,138 coupe, 1,382 convertible.

• The 24,056 G81 quantity included 2,385 3.08:1 ratio, 8,538 3.36:1 ratio, 3,119 3.55:1 ratio, 6,612 3.70:1 ratio, 3,310 4.11:1 ratio, and 91 4.56:1 ratio.

• The 2,401 M35 quantity was split 2,381 with 300hp, 20 with 390hp .

• RPO K66 (Transistor Ignition) was required purchase with L72, optional with L79 and L36.

• RPO K19 emission device was not limited to California use.

1966 Colors

CODE	EXTERIOR	QTY	SOFT TOP	WHEELS	INTERIORS
900	Tuxedo Black	1,190	Bk-W-Bg	Black	Bk-R-Bb-WB-S-Si-G-B
972	Ermine White	2,120	Bk-W-Bg	Black	Bk-R-Bb-WB-S-Si-G-B
974	Rally Red	3,366	Bk-W-Bg	Black	Bk-R
976	Nassau Blue	6,100	Bk-W-Bg	Black	Bk-Bb-WB-B
978	Laguna Blue	2,054	Bk-W-Bg	Black	Bk-Bb-B
980	Trophy Blue	1,463	Bk-W-Bg	Black	Bk-Bb-B
982	Mosport Green	2,311	Bk-W-Bg	Black	Bk-G
984	Sunfire Yellow	2,339	Bk-W-Bg	Black	Bk
986	Silver Pearl	2,967	Bk-W-Bg	Black	Bk-Si
988	Milano Maroon	3,799	Bk-W-Bg	Black	Bk-S

• Suggested interiors shown. Other combinations were possible.

• In 1966, 11 Corvettes were painted non-standard colors, or primer.

• Soft top colors were 7,259 black, 8,789 white, 411 beige.

Interior Codes: Std=Bk/V, 402=Bk/L, 407=R/V, 408=R/L, 414=Bb/V, 415=Bb/L, 418=B/V, 419=B/L, 420=S/V, 421=S/L, 426=Si/V, 427=Si/L, 430=G/V, 450=W+B/V.

Abbreviations: B=Blue, Bb=Bright Blue, Bg=Beige; Bk=Black, G=Green, L=Leather, R=Red, S=Saddle, Si=Silver, V=Vinyl, W=White, WB=White +Blue.

1967 Corvette

Production: 8,504 coupe, 14,436 convertible, 22,940 total.

1967 Numbers

Vehicle: 194377S100001 through 194377S122940
• For convertibles, fourth digit is a 6.

Suffix:

HD: 327ci, 350hp, mt, ar	IT: 427ci, 430hp (L88), mt
HE: 327ci, 300hp, mt	IU: 427ci, 435hp, mt, ah
HH: 327ci, 300hp, mt, ar	JA: 427ci, 435hp, mt, ar
HO: 327ci, 300hp, at	JC: 427ci, 400hp, mt
HP: 327ci, 350hp, mt, ac, ps	JD: 427ci, 400hp, at
HR: 327ci, 300hp, at, ar	JE: 427ci, 435hp, mt
HT: 327ci, 350hp, mt	JF: 427ci, 400hp, mt, ar
IL: 427ci, 390hp, mt	JG: 427ci, 400hp, at, ar
IM: 427ci, 390hp, mt, ar	JH: 427ci, 435hp, mt, ar, ah
IQ: 427ci, 390hp, at	KH: 327ci, 350hp, mt, ar, ac, ps
IR: 427ci, 390hp, at, ar	

Block:

3892657: 327ci	3869942: 427ci, ep, uu
3904351: 427ci	3916321: 427ci, lp, uu

Head:

3890462: 327ci, 300hp, 350hp	3904392: 427ci, 430hp, 435hp, ah
3904390: 427ci, 390hp, 400hp,ep	3909802: 427ci, 390hp, 400hp, lp
3904391: 427ci, 435hp, ih, ep	3919840: 427ci, 435hp, ih, lp

Carburetor:
Holley R3418A #3886091: 427ci, 430hp (L88)
Holley R3659A #3902353: 427ci, 400hp, 435hp, fc, rc
Holley R3660A #3902355: 427ci, 400hp,cc, mt; 435hp,cc
Holley R3810A #3906631: 327ci, 300hp, 350hp
Holley R3814A #3906635: 327ci, 300hp, 350hp, ar
Holley R3811A #3906633: 427ci, 390hp
Holley R3815A #3906637: 427ci, 390hp, ar
Holley R3888A #3909872: 427ci, 400hp, at, cc

Distributor:

1111117: 327ci, 300hp, at	1111141: 427ci, 390hp, ep
1111157: 327ci, 350hp, ig	1111240: 427ci, 430hp (L88)
1111194: 327ci, 300hp, mt	1111247: 427ci, 390hp, 400hp
1111196: 327ci, 350hp	1111248: 427ci, 390hp, 400hp, ig, ep
1111258: 427ci, 435hp, ig	1111294: 427ci, 390hp, 400hp, ig

Alternator:

1100693: 327ci, 427ci	1100696: 327ci, 427ci, ig
1100694: 327ci, 427ci, ac	1100750: 327ci, 427ci,ac,ig,uu

Ending Vehicle:

Sep 66: 102110	Jan 67: 109465	May 67: 119747
Oct 66: 102685	Feb 67: 112264	Jun 67: 122214
Nov 66: 104981	Mar 67: 115316	Jul 67: 122940
Dec 66: 107110	Apr 67: 117395	

Abbreviations: ac=air conditioning, ah=aluminum head, ar=air injection reactor, at=automatic transmission, cc=center carburetor, ci=cubic inch, ep=early production, fc=front carburetor, hp=horsepower, ig=transistor ignition, ih=iron head, lp=late production, mt=manual transmission, ps=power steering, rc=rear carburetor, uu=uncertain usage.

1967 Facts

• Last of the 1963-67 "mid years," 1967's exterior was the least adorned due to removal of trim, including hood script emblems and fender flags. Functional side fender vents were a new style with five slots

• Seats were a new design, and the parking brake handle was relocated from under the instrument panel to between the seats, a Corvette first. Inner door panels were similar to previous, but the lock buttons were located further forward and an attaching screw was added at the rear.

• Safety legislation required a modification of the knock-off wheel option. For 1967, it changed to a bolt-on, cast alloy style with a clip-on center cap to conceal the lug nuts. Rally wheels were no-cost standard equipment.

• The passenger hand hold above the glovebox was eliminated in 1967.

• The standard radio antenna for 1967 models had a fixed 31" mast, although adjustable antennas were available as dealer-installed options.

• A blue "GM Mark of Excellence" label was attached to the back of each 1967 door above the latch, a result of a GM quality awareness program.

1967 Options

RPO #	DESCRIPTION	QTY	RETAIL $
19437	Base Corvette Sport Coupe	8,504	$4,388.75
19467	Base Corvette Convertible	14,436	4,240.75
—	Genuine Leather Seats	1,601	79.00
A01	Soft Ray Tinted Glass, all windows	11,331	15.80
A02	Soft Ray Tinted Glass, windshield	6,558	10.55
A31	Power Windows	4,036	57.95
A82	Headrests	1,762	42.15
A85	Shoulder Belts	1,426	26.35
C07	Auxiliary Hardtop (for convertible)	6,880	231.75
C08	Vinyl Covering (for auxiliary hardtop)	1,966	52.70
C48	Heater and Defroster Deletion (credit)	35	-97.85
C60	Air Conditioning	3,788	412.90
F41	Special Front and Rear Suspension	2,198	36.90
G81	Positraction Rear Axle, all ratios	20,308	42.15
J50	Power Brakes	4,766	42.15
J56	Special Heavy Duty Brakes	267	342.30
K19	Air Injection Reactor	2,573	44.75
K66	Transistor Ignition System	5,759	73.75
L36	427ci, 390hp Engine	3,832	200.15
L68	427ci, 400hp Engine	2,101	305.50
L71	427ci, 435hp Engine	3,754	437.10
L79	327ci, 350hp Engine	6,375	105.35
L88	427ci, 430hp Engine	20	947.90
L89	Aluminum Cylinder Heads for L71	16	368.65
M20	4-Speed Manual Transmission	9,157	184.35
M21	4-Speed Man Trans, close ratio	11,015	184.35
M22	4-Speed Man Trans, close ratio, heavy duty	20	237.00
M35	Powerglide Automatic Transmission	2,324	194.35
N03	36 Gallon Fuel Tank (for coupe)	2	198.05
N11	Off Road Exhaust System	2,326	36.90
N14	Side Mount Exhaust System	4,209	131.65
N36	Telescopic Steering Column	2,415	42.15
N40	Power Steering	5,747	94.80
N89	Cast Alluminum Bolt-On Wheels (5)	720	263.30
P92	Whitewall Tires, 7.75x15	13,445	31.35
QB1	Redline Tires, 7.75x15	4,230	46.65
U15	Speed Warning Indicator	2,108	10.55
U69	AM-FM Radio	22,193	172.75

• A 327ci, 300hp engine, 3-speed manual transmission, vinyl interior trim, and soft top (convertible) were included in the base price.

• The 6,880 C07 quantity included 895 in lieu of soft tops at no extra cost.

• The 3,788 C60 quantity was split 2,235 coupe, 1,553 convertible.

• The 20,308 G81 quantity included 1,981 3.08:1 ratio, 8,189 3.36:1 ratio, 2,410 3.55:1 ratio, 5,662 3.70:1 ratio, 2,061 4.11:1 ratio and 5 4.56:1 ratio.

• The 2,324 M35 quantity was split 1,725 with 300hp engines, 392 with 390hp engines, and 207 with 400hp engines.

1967 Colors

CODE	EXTERIOR	QTY	SOFT TOP	WHEELS	INTERIORS
900	Tuxedo Black	815	Bk-W-Tb	Silver	Bk-R-Bb-S-W-Tb-G
972	Ermine White	1,423	Bk-W-Tb	Silver	Bk-R-Bb-S-W-Tb-G
974	Rally Red	2,341	Bk-W-Tb	Silver	Bk-R-W
976	Marina Blue	3,840	Bk-W-Tb	Silver	Bk-Bb-W
977	Lynndale Blue	1,381	Bk-W-Tb	Silver	Bk-W-Tb
980	Elkhart Blue	1,096	Bk-W-Tb	Silver	Bk-Tb
983	Goodwood Green	4,293	Bk-W-Tb	Silver	Bk-S-W-G
984	Sunfire Yellow	2,325	Bk-W-Tb	Silver	Bk-W
986	Silver Pearl	1,952	Bk-W-Tb	Silver	Bk-Tb
988	Marlboro Maroon	3,464	Bk-W-Tb	Silver	Bk-W-S

• Suggested interiors shown. Other combinations were possible.

• In 1967, 10 Corvettes had non-standard paint, or primer.

• Soft top colors were 7,030 black, 5,900 white, 611 teal blue.

Interiors: Std=Bk/V, 402=Bk/L, 407=R/V, 408=R/L, 414=Bb/V, 415=Bb/L, 418=Tb/V, 419=Tb/L, 420=S/V, 421=S/L, 430=G/V, 450=W+Bb/V, 455=W+Bk/V.

Abbreviations: Bb=Bright Blue, Bk=Black, G=Green, L=Leather, R=Red, S=Saddle, Tb=Teal Blue, V=Vinyl, W=White.

1968 Corvette

Production: 9,936 coupe, 18,630 convertible, 28,566 total.

1968 Numbers

Vehicle: 194378S400001 through 194378S428566
    • For convertibles, fourth digit is a 6.

Suffix:
HE: 327ci, 300hp, mt
HO: 327ci, 300hp, at
HP: 327ci, 350hp, mt, ac, ps
HT: 327ci, 350hp, mt
IL: 427ci, 390hp, mt
IM: 427, 400hp, mt

IO: 427ci, 400hp, at
IQ: 427ci, 390hp, at
IR: 427ci, 435hp, mt
IT: 427ci, 430hp (L88), mt, ah
IU: 427ci, 435hp, mt, ah

Block: 3914678: 327ci, 300hp, 350hp
3916321: 427ci, 390hp, 400hp, 430hp, 435hp, fd
3935439: 427ci, 390hp, 400hp, 430hp, 435hp, sd

Head:
3917215: 427ci, 390hp, 400hp
3917291: 327ci, 300hp, 350hp
3917292: 327ci, 350hp
3919840: 427ci, 435hp, ih
3919842: 427ci, 430hp, 435hp, ah

Carb: Rochester Q-jet #7028207: 327ci, 300hp, mt
Rochester Q-jet #7028208: 327ci, 300hp, at
Rochester Q-jet #7028209: 427ci, 390hp, mt
Rochester Q-jet #7028216: 427ci, 390hp, at
Rochester Q-jet #7028219: 327ci, 350hp, mt
Holley R3659A #3902353: 427ci, 400hp, 435hp, fc, rc
Holley R4054A #3925519: 427ci, 430hp (L88)
Holley R4055A #3925517: 427ci, 400hp, cc, mt; 435hp, cc, fd
Holley R4055-1A #3940929: 427ci, 400hp, cc, mt; 435hp, cc, sd
Holley R4056A #3902516: 427ci, 400hp, cc, at, fd
Holley R4056-1A #3940930: 427ci, 400hp, cc, at, sd

Distributor:
1111194: 327ci, 300hp
1111293: 427ci, 390hp,400hp
1111294: 427ci, 390hp,400hp,ig
1111295: 427ci, 430hp,ig
1111296: 427ci, 435hp,ig
1111438: 327ci, 350hp
1111441: 327ci, 350hp,ig,fd
1111475: 327ci, 350hp,ig,fd
1111477: 327ci, 350hp,sd

Alternator: 1100693: 300hp, 350hp, 390hp, 400hp
1100696: 350hp, 390hp, 400hp, 430hp, 435hp, ig
1100750: All with ac

Ending Vehicle:
Sep 67: 400905
Oct 67: 403410
Nov 67: 405682
Dec 67: 407922
Jan 68: 410386
Feb 68: 412647
Mar 68: 415000
Apr 68: 417676
May 68: 420928
Jun 68: 423978
Aug 68: 428566

Abbreviations: ac=air conditioning, ah=aluminum heads, at=automatic transmission, cc=center carb, ci=cubic inch, fc=front carb, fd=first design, hp=horsepower, ig=transistor ignition, ih=iron head, mt=manual transmission, ps=power steering, rc=rear carb, sd=second design.

1968 Facts

• The wheelbase remained the same and much of the chassis carried over, but exterior and interior were extensively redesigned. Coupes featured removable roof T-tops and removable rear window.

• Automatic transmissions changed for 1968 from the two-speed Powerglide to the three-speed Turbo Hydra-Matic.

• Hidden headlights were operated by vacuum and moved up into position, rather than being rotated electrically as with 1963-1967 models.

• A light monitoring system was standard and utilized fiber optics to display functioning lamps on a central console display.

• The battery moved to a compartment behind the seats to improve weight distribution and to free underhood space.

• Windshield wipers were hidden under a vacuum-operated panel.

• Side vent windows were gone, but Astro Ventilation was new. This was the last with an ignition switch on the instrument panel until 1997.

• 1968's door opening mechanism, a spring-loaded finger plate and separate release button, was unique to the year.

1968 Options

RPO #	DESCRIPTION	QTY	RETAIL$
19437	Base Corvette Sport Coupe	9,936	$4,663.00
19467	Base Corvette Convertible	18,630	4,320.00
—	Genuine Leather Seats	2,429	79.00
A01	Soft Ray Tinted Glass, all windows	17,635	15.80
A02	Soft Ray Tinted Glass, windshield	5,509	10.55
A31	Power Windows	7,065	57.95
A82	Headrests	3,197	42.15
A85	Custom Shoulder Belts (std with coupe)	350	26.35
C07	Auxiliary Hardtop (for convertible)	8,735	231.75
C08	Vinyl Covering (for auxiliary hardtop)	3,050	52.70
C50	Rear Window Defroster	693	31.60
C60	Air Conditioning	5,664	412.90
F41	Special Front and Rear Suspension	1,758	36.90
G81	Positraction Rear Axle, all ratios	27,008	46.35
J50	Power Brakes	9,559	42.15
J56	Special Heavy Duty Brakes	81	384.45
K66	Transistor Ignition System	5,457	73.75
L36	427ci, 390hp Engine	7,717	200.15
L68	427ci, 400hp Engine	1,932	305.50
L71	427ci, 435hp Engine	2,898	437.10
L79	327ci, 350hp Engine	9,440	105.35
L88	427ci, 430hp Engine	80	947.90
L89	Aluminum Cylinder Heads with L71	624	805.75
M20	4-Speed Manual Transmission	10,760	184.35
M21	4-Speed Man Trans, close ratio	12,337	184.35
M22	4-Speed Man Trans, close ratio, heavy duty	80	263.30
M40	Turbo Hydra-Matic Automatic Transmission	5,063	226.45
N11	Off Road Exhaust System	4,695	36.90
N36	Telescopic Steering Column	6,477	42.15
N40	Power Steering	12,364	94.80
P01	Bright Metal Wheel Cover	8,971	57.95
PT6	Red Stripe Tires, F70x15, nylon	11,686	31.30
PT7	White Stripe Tires, F70x15, nylon	9,692	31.30
UA6	Alarm System	388	26.35
U15	Speed Warning Indicator	3,453	10.55
U69	AM-FM Radio	24,609	172.75
U79	AM-FM Radio, stereo	3,311	278.10

• A 327ci, 300hp engine, 3-speed manual transmission, vinyl interior trim, and soft top (convertible) were included in the base price.

• RPO A85 (Shoulder Belts) were included with all coupes at no charge. Quantity shown was for convertible installations which were added cost.

• The M40 Turbo Hydra-Matic transmission cost $226.45 when combined with the 327ci, 300hp engine, but was $237.00 when combined with the 427ci, 390hp or 400hp engines.

1968 Colors

CODE	EXTERIOR	QTY	SOFT TOP	INTERIORS
900	Tuxedo Black	708	Bk-W-Bg	Bk-Db-Do-Gu-Mb-R-To
972	Polar White	1,868	Bk-W-Bg	Bk-Db-Do-Gu-Mb-R-To
974	Rally Red	2,918	Bk-W-Bg	Bk-R
976	LeMans Blue	4,722	Bk-W-Bg	Bk-Mb-Db
978	International Blue	2,473	Bk-W-Bg	Bk-Mb-Db
983	British Green	4,779	Bk-W-Bg	Bk
984	Safari Yellow	3,133	Bk-W-Bg	Bk
986	Silverstone Silver	3,435	Bk-W-Bg	Bk-Db-Gu
988	Cordovan Maroon	1,155	Bk-W-Bg	Bk
992	Corvette Bronze	3,374	Bk-W-Bg	Bk-Do-To

• Suggested interiors shown. Other combinations were possible.

• Starting in 1968, the body paint and trim plate was riveted to the driver's door hinge pillar.

• Wheels were 7"wide (unique to 1968). All were painted silver.

• In 1968, one Corvette had non-standard paint, believed to be pink.

Interior Codes: Std=Bk/V, 402=Bk/L, 407=R/V, 408=R/L, 411=Db/V, 414=Mb/V, 415=Mb/L, 425=Do/V, 426=Do/L, 435=To/V, 436=To/L, 442=Gu/V.

Abbreviations: Bg=Beige, Bk=Black, Db=Dark Blue, Do=Dark Orange, Gu=Gunmetal, L=Leather, Mb=Medium Blue, R=Red, To=Tobacco, V=Vinyl, W=White.

1969 Corvette

Production: 22,129 coupe, 16,633 convertible, 38,762 total.

1969 Numbers

Vehicle: 194379S700001 through 194379S738762
  • For convertibles, fourth digit is a 6.

Suffix:
GC: 350ci, 350hp, mt, ig
GD: 350ci, 350hp, mt, ac, ig
HW: 350ci, 350hp, mt
HX: 350ci, 350hp, mt, ac
HY: 350ci, 300hp, mt
HZ: 350ci, 300hp, at
LL: 427ci, 390hp, at
LM: 427ci, 390hp, mt
LN: 427ci, 400hp, at
LO: 427ci, 430hp, mt (L88)
LP: 427ci, 435hp, mt, ah
LQ: 427ci, 400hp, mt
LR: 427ci, 435hp, mt

LT: 427ci, 435hp, mt, hc
LU: 427ci, 435hp, mt, ah, hc
LV: 427ci, 430hp(L88), at
LW: 427ci, 435hp, at, ah
LX: 427ci, 435hp, at
ME: 427ci, 430hp(ZL1), mt
MG: 427ci, 430hp(ZL1), at
MH: 427ci, 390hp, mt, ig
MI: 427ci, 390hp, at, ig
MJ: 427ci, 400hp, at, ig
MK: 427ci, 400hp, mt, ig
MR: 427ci, 430hp(L88), mt
MS: 427ci, uu

Block:
3932386, 3932388: 350ci, 300hp, 350hp
3935439: 427ci, 390hp, 400hp, 430hp, 435hp
3946052: 427ci, 430hp (ZL1)
3955270, 3963512: 427ci, 390hp, 400hp, 430hp, 435hp
3956618, 3970010: 350ci, 300hp, 350hp

Head:
3919840: 427ci, 435hp, ih
3919842: 427ci, 435hp, ah
3927186: 350ci, 300hp, 350hp
3927187: 350ci, 350hp

3931063: 427ci, 390hp, 400hp
3946074: 427ci, 430hp, ah
3947041: 350ci, 300hp

Carb:
Rochester Q-jet #7029202: 350ci, 300hp, at
Rochester Q-jet #7029203: 350ci, 300hp, mt
Rochester Q-jet #7029204: 427ci, 390hp, at
Rochester Q-jet #7029207: 350ci, 350hp, mt
Rochester Q-jet #7029215: 427ci, 390hp, mt
Holley R3659A #3902353: 427ci, 400hp, fc, rc; 435hp, fc, rc
Holley R4055-1A #3940929: 427ci, 400hp, cc, mt; 435hp, cc
Holley R4056-1A #3940930: 427ci, 400hp, cc, at
Holley R4054A #3925519: 427ci, 430hp, fd
Holley R4296A #3955205: 427ci, 430hp, sd

Distributor:
1111490: 350ci, 300hp
1111491: 350ci, 350hp, ig
1111493: 350ci, 350hp
1111927: 427ci, 430hp, ig

1111926: 427ci, 390hp, 400hp
1111928: 427ci, 435hp, ig
1111954: 427ci, 390hp, 400hp, ig

Alternator:
1100825: ac and / or ig
1100833: 427ci, 390hp, 400hp
1100859: 350ci, 300hp, 350hp

1100882: 427ci, 430hp,435hp,ig
1100884: 350ci, 300hp, ac (uu)
1100900: (uu)

Ending Vehicle:

	Dec 68: 711742	Apr 69: 721315	Sep 69: 730963
Sep 68: 703041	Jan 69: 714695	Jun 69: 723374	Oct 69: 734067
Oct 68: 706272	Feb 69: 717571	Jul 69: 725875	Nov 69: 736798
Nov 68: 709159	Mar 69: 720543	Aug 69: 728107	Dec 69: 738762

Abbreviations: ac=air conditioning, ah=aluminum heads, at=automatic transmission, cc=center carburetor, ci=cubic inch, fd=first design, fc=front carburetor, hc=heavy-duty clutch, hp=horsepower, ig=transistor ignition, ih=iron head, mt=manual transmission, ps=power steering, rc=rear carburetor, sd=second design, uu=uncertain usage.

1969 Facts

• Appearance was similar to 1968 with small differences. A "Stingray" (one word) script was added to front fenders. The exterior door opening mechanism was designed into what was simply a finger hold in 1968, eliminating the separate release button. Optional side mount exhausts and side fender vent trim also differentiated 1969 from 1968.

• It was the first year for 350ci displacement engines, a steering column ignition switch, 8" wheel width, and headlight washers.

• The steering wheel diameter was reduced from 16" to 15". Map pockets were added to the passenger side dash area.

1969 Options

RPO #	DESCRIPTION	QTY	RETAIL $
19437	Base Corvette Sport Coupe	22,129	$4,781.00
19467	Base Corvette Convertible	16,633	4,438.00
—	Genuine Leather Seats	3,729	79.00
A01	Soft Ray Tinted Glass, all windows	31,270	16.90
A31	Power Windows	9,816	63.20
A82	Headrests	38,762	17.95
A85	Custom Shoulder Belts (std with coupe)	600	42.15
C07	Auxiliary Hardtop (for convertible)	7,878	252.80
C08	Vinyl Covering (for auxiliary hardtop)	3,266	57.95
C50	Rear Window Defroster	2,485	32.65
C60	Air Conditioning	11,859	428.70
F41	Special Front and Rear Suspension	1,661	36.90
G81	Positraction Rear Axle, all ratios	36,965	46.35
J50	Power Brakes	16,876	42.15
J56	Special Heavy Duty Brakes	115	384.45
K05	Engine Block Heater	824	10.55
K66	Transistor Ignition System	5,702	81.10
L36	427ci, 390hp Engine	10,531	221.20
L46	350ci, 350hp Engine	12,846	131.65
L68	427ci, 400hp Engine	2,072	326.55
L71	427ci, 435hp Engine	2,722	437.10
L88	427ci, 430hp Engine	116	1,032.15
L89	Aluminum Cylinder Heads with L71	390	832.05
MA6	Heavy Duty Clutch	102	79.00
M20	4-Speed Manual Transmission	16,507	184.80
M21	4-Speed Man Trans, close ratio	13,741	184.80
M22	4-Speed Man Trans, close ratio, heavy duty	101	290.40
M40	Turbo Hydra-Matic Automatic Transmission	8,161	221.80
N14	Side Mount Exhaust System	4,355	147.45
N37	Tilt-Telescopic Steering Column	10,325	84.30
N40	Power Steering	22,866	105.35
P02	Deluxe Wheel Covers	8,073	57.95
PT6	Red Stripe Tires, F70x15, nylon	5,210	31.30
PT7	White Stripe Tires, F70x15, nylon	21,379	31.30
PU9	White Letter Tires, F70x15, nylon	2,398	33.15
TJ2	Front Fender Louver Trim	11,962	21.10
UA6	Alarm System	12,436	26.35
U15	Speed Warning Indicator	3,561	11.60
U69	AM-FM Radio	33,871	172.75
U79	AM-FM Radio, stereo	4,114	278.10
ZL1	Special L88 (aluminum block)	2	4,718.35

• A 350ci, 300hp engine, 3-speed manual transmission, vinyl interior, and soft top (conv) or T-tops (cpe) were included in the base.

•RPO A82 was a required RPO early, then standard after January 1, 1969. However, GM records indicate it could be deleted for credit.

• RPO M40 cost $290.40 with L71, L88 or L89.

• RPO ZL1 was all aluminum 427ci with required L88 options K66, F41, J56 and G81. Radio and air conditioning not available with ZL1.

1969 Colors

CODE	EXTERIOR	SOFT TOP	WHEELS	INTERIORS
900	Tuxedo Black	Bk-W-Bg	Silver	Bk-Bb-G-Gu-R-S
972	Can-Am White	Bk-W-Bg	Silver	Bk-Bb-G-Gu-R-S
974	Monza Red	Bk-W-Bg	Silver	Bk-R-S
976	LeMans Blue	Bk-W-Bg	Silver	Bk-Bb
980	Riverside Gold	Bk-W-Bg	Silver	Bk-S
983	Fathom Green	Bk-W-Bg	Silver	Bk-G-S
984	Daytona Yellow	Bk-W-Bg	Silver	Bk
986	Cortez Silver	Bk-W-Bg	Silver	Bk-Bb-G-Gu-R-S
988	Burgundy	Bk-W-Bg	Silver	Bk-S
990	Monaco Orange	Bk-W-Bg	Silver	Bk

• Suggested interiors shown. Other combinations were possible.

• 1969 production continued through December, 1969. Interior color quantities currently available are through August 31, 1969: 15,347 Bk/V, 1,649 B/L, 1,531 R/V, 181 R/L, 3,527 Bb/V, 242 Bb/L, 606 Gu/V, 96 Gu/L, 2,956 S/V, 641 S/L, 1,217 G/V, 115 G/L.

Interior Codes: ZQ4 or std=Bk/V, 402=Bk/L, 407=R/V, 408=R/L, 411=Bb/V, 412=Bb/L, 416=Gu/V, 417=Gu/L, 420=S/V, 421=S/L, 427=G/V, 428=G/L.

Abbreviations: Bb=Bright Blue, Bg=Beige, Bk=Black, G=Green, Gu=Gunmetal, L=Leather, R=Red, S=Saddle, V=Vinyl, W=White.

1970 Corvette

Production: 10,668 coupe, 6,648 convertible, 17,316 total.

1970 Numbers

Vehicle: 194370S400001 through 194370S417316
- For convertibles, fourth digit is a 6.

Suffix:

CGW: 454ci, 390hp, at	CTM: 350ci, 300hp, at
CRJ: 454ci, 390hp, at, ig	CTN: 350ci, 350hp, mt
CRI: 454ci, 390hp, mt, ig	CTO: 350ci, 350hp, mt, ac
CTD: 350ci, 300hp, mt	CTP: 350ci, 350hp, mt, ig
CTG: 350ci, 300hp, at	CTQ: 350ci, 350hp, mt, ac, ig
CTH: 350ci, 350hp, mt	CTR: 350ci, 370hp, mt, ig
CTJ: 350ci, 350hp, mt, ac	CTU: 350ci, 370hp, mt, ig
CTK: 350ci, 370hp, mt, ig	CTV: 350ci, 370hp (ZR1), mt
CTL: 350ci, 300hp, mt	CZU: 454ci, 390hp, mt

- Some early 1970 models may have 1969 engine codes.

Block: 3970010: 350ci, 300hp, 350hp, 370hp
      3963512: 454ci, 390hp

Head: 3927186: 350ci, 300hp, 350hp, 370hp   3964290: 454ci, 390hp
     3927187: 350ci, 350hp   3973414: 350ci, 370hp

Carburetor: Rochester Q-jet #7040202: 350ci, 300hp, at, fd
                Rochester Q-jet #7040203: 350ci, 300hp, mt, fd
                Rochester Q-jet #7040204: 454ci, 390hp, at
                Rochester Q-jet #7040205: 454ci, 390hp, mt
                Rochester Q-jet #7040207: 350ci, 350hp, mt
                Rochester Q-jet #7040212: 350ci, 300hp, at, sd
                Rochester Q-jet #7040213: 350ci, 300hp, mt, sd
                Rochester Q-jet #7040502: 350ci, 300hp, at, ec
                Rochester Q-jet #7040503: 350ci, 300hp, mt, ec, fd
                Rochester Q-jet #7040504: 454ci, 390hp, at, ec
                Rochester Q-jet #7040505: 454ci, 390hp, mt, ec
                Rochester Q-jet #7040507: 350ci, 350hp, mt, ec
                Rochester Q-jet #7040513: 350ci, 300hp, mt, ec, sd
                Holley R4489A #3972123: 350ci, 370hp, mt, ec
                Holley R4555A #3972121: 350ci, 370hp, mt

Distributor: 1111464: 454ci, 390hp   1111493: 350ci, 350hp, ep
             1111490: 350ci, 300hp, ep   1112020: 350ci, 300hp
             1111491: 350ci, 370hp, ig   1112021: 350ci, 350hp

Alternator: 1100884: 350ci, 370hp or ac(all)   1100901: 350ci, 300hp
            1100900: 350ci(350hp), 454ci(390hp)

Ending Vehicle: Jan 70: 402261   Apr 70: 408314   Jul 70: 417316
               Feb 70: 405183   May 70: 410652
               Mar 70: 407977   Jun 70: 413829

Abbreviations: ac=air conditioning, at=automatic transmission, ci=cubic inch, ec=evaporative emission control, ep=early production, fd=first design, hp=horsepower, ig=transistor ignition, mt=manual transmission, sd=second design.

1970 Facts

- Body design for 1970 was updated with fender flairs to reduce wheel-thrown debris damage, a problem with 1968-69 models. New fender louvers replaced the four vertical slots of the 1968-69 models. Exhaust tips were rectangular, replacing the round style of earlier years. Front grills were square mesh. Parking lamps, located at the grills' outer edge, had clear lenses with amber bulbs.

- Interiors in 1970 were refined, including redesigned seats for additional headroom and easier access to rear storage. For the first time, deluxe interiors were optional which included leather seating, wood graining for console and door panels, and special carpeting.

- Big block engine displacement increased from 427ci to 454ci. Small block displacement remained 350ci, but a new LT1 engine option had solid lifters and 370hp.

- Due to extended 1969 production, 1970 production did not begin until January, 1970. Total build for 1970 was 17,316, lowest since 1962.

1970 Options

RPO #	DESCRIPTION	QTY	RETAIL $
19437	Base Corvette Sport Coupe	10,668	$5,192.00
19467	Base Corvette Convertible	6,648	4,849.00
—	Custom Interior Trim	3,191	158.00
A31	Power Windows	4,813	63.20
A85	Custom Shoulder Belts (std with coupe)	475	42.15
C07	Auxiliary Hardtop (for convertible)	2,556	273.85
C08	Vinyl Covering (for auxiliary hardtop)	832	63.20
C50	Rear Window Defroster	1,281	36.90
C60	Air Conditioning	6,659	447.65
G81	Optional Rear Axle Ratio	2,862	12.65
J50	Power Brakes	8,984	47.40
L46	350ci, 350hp Engine	4,910	158.00
LS5	454ci, 390hp Engine	4,473	289.65
LT1	350ci, 370hp Engine	1,287	447.60
M21	4-Speed Man Trans, close ratio	4,383	0.00
M22	4-Speed Man Trans, close ratio, heavy duty	25	95.00
M40	Turbo Hydra-Matic Automatic Transmission	5,102	0.00
NA9	California Emissions	1,758	36.90
N37	Tilt-Telescopic Steering Column	5,803	84.30
N40	Power Steering	11,907	105.35
P02	Deluxe Wheel Covers	3,467	57.95
PT7	White Stripe Tires, F70x15, nylon	6,589	31.30
PU9	White Letter Tires, F70x15, nylon	7,985	33.15
T60	Heavy Duty Battery (std with LS5)	165	15.80
UA6	Alarm System	6,727	31.60
U69	AM-FM Radio	14,529	172.75
U79	AM-FM Radio, stereo	2,462	278.10
ZR1	Special Purpose Engine Package	25	968.95

• A 350ci, 300hp engine, 4-speed wide ratio manual transmission, vinyl interior trim, soft top (conv) or T-tops (cpe) were included in the base.

• Previously optional tinted glass, Positraction axle, and 4-speed wide ratio manual transmission were also included in the 1970 base price.

• ZR1 included LT1 engine, M22 transmission, heavy duty power brakes, transistor ignition, special aluminum radiator, and special springs, shocks and front and rear stabilizer bars. RPOs A31, C50, C60, N40, P02, UA6, U69 and U79 were not available. ZR1's also had metal fan shrouds.

• An LS7 rated at 460hp (or 465hp) was planned and appeared in order guides. It was to be the big-block version of the LT1, including aluminum heads. A ZR2 package with the LS7 was also planned. Suffix for the LS7 was assigned as CZL for manual, CZN for automatic. The LS7 and ZR2 were cancelled and it is thought none were delivered to retail customers.

• Custom interior included leather seat trim, wood-grain accents and carpet trim on door panels, wood-grain accents on console, and cut-pile carpeting.

1970 Colors

CODE	EXTERIOR	SOFT TOP	WHEELS	INTERIORS
972	Classic White	Bk-W-Bg	Silver	B-Bk-Br-G-R-S
974	Monza Red	Bk-W-Bg	Silver	Bk-Br-R-S
975	Marlboro Maroon	Bk-W-Bg	Silver	Bk-Br-S
976	Mulsanne Blue	Bk-W-Bg	Silver	B-Bk
979	Bridgehampton Blue	Bk-W-Bg	Silver	B-Bk
982	Donnybrooke Green	Bk-W-Bg	Silver	Bk-Br-G-S
984	Daytona Yellow	Bk-W-Bg	Silver	Bk-G
986	Cortez Silver	Bk-W-Bg	Silver	B-Bk-Br-G-R-S
991	Ontario Orange	Bk-W-Bg	Silver	Bk-S
992	Laguna Gray	Bk-W-Bg	Silver	B-Bk-Br-G-R-S
993	Corvette Bronze	Bk-W-Bg	Silver	Bk

• Suggested interiors shown. Other combinations were possible.

• Custom interiors available only in black or saddle colors.

• Exterior color quantities are not available from Chevrolet at this time, but statistics based on 957 cars compiled by the 1970 Corvette Registry (www.1970CorvetteRegistry.Org) show code 972=8.3%, 974=13.6%, 975=7.9%, 976=10.2%, 979=9%, 982=19.4%, 984=10.4%, 986=6.9%, 991=0%, 992=7.5%, 993=6.7%.

Interior Codes: 400=Bk/V, 403=Bk/L, 407=R/V, 411=B/V, 414=Br/V, 418=S/V, 422=G/V, 424=S/L.

Abbreviations: B=Blue, Bk=Black, Bg=Beige, Br=Brown, G=Green, L=Leather, R=Red, S=Saddle, V=Vinyl, W=White.

1971 Corvette

Production: 14,680 coupe, 7,121 convertible, 21,801 total.

1971 Numbers

Vehicle: 194371S100001 through 194371S121801
• For convertibles, fourth digit is a 6.

Suffix:
CGT: 350ci, 270hp, at, uu	CPH: 454ci, 365hp, mt
CGY: 350ci, 330hp(ZR1), mt	CPJ: 454ci, 365hp, at
CGZ: 350ci, 330hp(LT1), mt	CPW: 454ci, 425hp, mt, ah
CJK: 350ci, 270hp, at	CPX: 454ci, 425hp, at, ah
CJL: 350ci, 270hp, mt	

Block: 3970010: 350ci, 270hp, 330hp
3963512: 454ci, 365hp, 425hp

Head: 3946074: 454ci, 425hp   3993820: 454ci, 365hp
3973487: 350ci, 270hp, 330hp   3994026: 454ci, 425hp (uu)

Carburetor: Rochester Q-jet #7041204: 454ci, 365hp, at
Rochester Q-jet #7041205: 454ci, 365hp, mt
Rochester Q-jet #7041212: 350ci, 270hp, at
Rochester Q-jet #7041213: 350ci, 270hp, mt
Holley R4801A #3989021: 350ci, 330hp, mt
Holley R4802A #3986195: 454ci, 425hp, mt
Holley R4803A #3986196: 454ci, 425hp, at

Distributor: 1112038: 350ci, 330hp, ig   1112053: 454ci, 425hp, at, ig
1112050: 350ci, 270hp   1112076: 454ci, 425hp, mt, ig
1112051: 454ci, 365hp

Alternator: 1100543: 454ci, 365hp, 425hp   1100544: All with ac
1100950: 350ci, 270hp, 330hp

Ending Vehicle:
Aug 70: 101212	Jan 71: 108230	May 71: 118223
Sep 70: 102226	Feb 71: 110886	Jun 71: 120686
Nov 70: 102675	Mar 71: 113626	Jul 71: 121801
Dec 70: 105269	Apr 71: 115983	

Abbreviations: ac=air conditioning, ah=aluminum heads, at=automatic transmission, ci=cubic inch, hp=horsepower, ig=transistor ignition, mt=manual transmission, uu=uncertain usage.

1971 Facts

• The 1971 Corvette was one of the least-changed in appearance. A labor dispute in May 1969 caused 1969 production to run long, shortening normal 1970 production by over four months. Chevrolet then treated 1971 Corvette production as an extension of 1970. Plus, GM directed its divisions to reduce octane requirements in 1971 engines, an effort which carried a higher priority than appearance changes.

• 1971 engines were detuned variants of 1970 engines. General Motors' intent in reducing octane requirements to 91 (research) was to give oil companies phase-in time for the unleaded fuels needed for future use of catalytic converters.

• The 454ci LS6 engine with 425hp was a detuned version of 1970's planned but cancelled 460hp LS7. LS6 was designed to operate on low-lead fuel, but since a comparable engine was not available in 1970, it won the horsepower race for the two years despite its lower octane appetite. It featured aluminum heads and could be ordered with an automatic transmission, although not when combined with the ZR2 package.

• Production specifications called for amber parking light lenses, but most 1971s had clear lenses with amber bulbs as in 1970.

• The fuel filler door hardware was redesigned in 1971 for easier access.

• The 1971 Corvette was the last model to feature the fiber-optics light monitoring system, although late production 1971s may be missing the fiber optics. It has been reported that the cost savings permitted inclusion of the previously optional antitheft alarm system as standard equipment the following year.

• Seating pleats ran fore and aft for vinyl seats and for most leather seats, but later production 1971 leather seats had pleats running side to side.

1971 Options

RPO #	DESCRIPTION	QTY	RETAIL $
19437	Base Corvette Sport Coupe	14,680	$5,496.00
19467	Base Corvette Convertible	7,121	5,259.00
—	Custom Interior Trim	2,602	158.00
A31	Power Windows	6,192	79.00
A85	Custom Shoulder Belts (std with coupe)	677	42.00
C07	Auxiliary Hardtop (for convertible)	2,619	274.00
C08	Vinyl Covering (for auxiliary hardtop)	832	63.00
C50	Rear Window Defroster	1,598	42.00
C60	Air Conditioning	11,481	459.00
ZQ1	Optional Rear Axle Ratio	2,395	13.00
J50	Power Brakes	13,558	47.00
LS5	454ci, 365hp Engine	5,097	295.00
LS6	454ci, 425hp Engine	188	1,221.00
LT1	350ci, 330hp Engine	1,949	483.00
M21	4-Speed Man Trans, close ratio	2,387	0.00
M22	4-Speed Man Trans, close ratio, heavy duty	130	100.00
M40	Turbo Hydra-Matic Automatic Transmission	10,060	0.00
N37	Tilt-Telescopic Steering Column	8,130	84.30
N40	Power Steering	17,904	115.90
P02	Deluxe Wheel Covers	3,007	63.00
PT7	White Stripe Tires, F70x15, nylon	6,711	28.00
PU9	White Letter Tires, F70x15, nylon	12,449	42.00
T60	Heavy Duty Battery (std with LS5, LS6)	1,455	15.80
UA6	Alarm System	8,501	31.60
U69	AM-FM Radio	18,078	178.00
U79	AM-FM Radio, stereo	3,431	283.00
ZR1	Special Purpose LT1 Engine Package	8	1,010.00
ZR2	Special Purpose LS6 Engine Package	12	1,747.00

• A 350ci, 270hp engine, 4-speed wide-ratio manual transmission, vinyl interior trim, and soft top (conv) or T-tops were included in the base.

• RPO ZR1 (Special Purpose LT1 Engine Package) included the LT1 engine, M22 transmission, heavy-duty power brakes, transistor ignition, special aluminum radiator, and special springs, shocks, and front and rear stabilizer bars (ZR1s have appeared with and without rear stabilizers). ZR1s also had metal fan shrouds. RPOs A31, C50, C60, N40, P02, UA6, U69 and U79 were not available with ZR1.

• RPO ZR2 (Special Purpose LS6 Engine Package) had the same content and ordering restrictions as the ZR1 package except ZR2 included the 425hp, 454ci LS6 engine. The LS6 engines could be combined with automatic transmission, but not when part of the ZR2 package.

• Custom interior included leather seat trim, wood-grain accents and lower carpet trim on interior door panels, wood-grain accents on console, and special cut-pile carpeting.

• M40 was no cost with the base 350ci, 270hp engine, but cost $100.35 with LS5 or LS6. It was not available with LT1, ZR1 or ZR2.

1971 Colors

CODE	EXTERIOR	QTY	SOFTTOP	WHEELS	INTERIORS
905	Nevada Silver	1,177	Bk-W	Silver	Bk-Db-Dg-R
912	Sunflower Yellow	1,177	Bk-W	Silver	Bk-Dg-S
972	Classic White	1,875	Bk-W	Silver	Bk-Db-Dg-R-S
973	Mille Miglia Red	2,180	Bk-W	Silver	Bk-R
976	Mulsanne Blue	2,465	Bk-W	Silver	Bk-Db
979	Bridgehampton Blue	1,417	Bk-W	Silver	Bk-Db
983	Brands Hatch Green	3,445	Bk-W	Silver	Bk-Dg
987	Ontario Orange	2,269	Bk-W	Silver	Bk-Dg-S
988	Steel Cities Gray	1,591	Bk-W	Silver	Bk-S
989	War Bonnet Yellow	3,706	Bk-W	Silver	Bk-Dg-S

• Suggested interiors shown. Other combinations were possible.

• For reasons unknown, total of colors is 499 less than total production.

• Custom interiors, which included leather seating, were available only in black or saddle colors.

Interior Codes: std/400=Bk/V, 403=Bk/L, 407=R/V, 412=Db/V, 417=S/V, 420=S/L, 423=Dg/V.

Abbreviations: Bk=Black, Db=Dark Blue, Dg=Dark Green, L=Leather, R=Red, S=Saddle, V=Vinyl, W=White.

1972 Corvette

1972 Numbers

Vehicle: 1Z37K2S500001 through 1Z37K2S527004
- For convertibles, third digit is a 6.
- Fifth digit varies as follows: K=350ci, 200hp; L=350ci, 255hp; W=454ci, 270hp.

Suffix:
CDH: 350ci, 200hp, mt, ec	CPH: 454ci, 270hp, mt
CDJ: 350ci, 200hp, at, ec	CPJ: 454ci, 270hp, at, uu
CKW: 350ci, 200hp, mt	CRS: 350ci, 255hp, ar, at
CKX: 350ci, 200hp, at	CRT: 350ci, 255hp, mt, ar, uu
CKY: 350ci, 255hp, mt	CSR: 454ci, 270hp, mt, ar
CKZ: 350ci, 255hp(ZR1), mt	CSS: 454ci, 270hp, at, ar

Block: 3970010: 350ci, 200hp, 255hp   3999289: 454, 270hp
3970014: 350ci, 200hp, 255hp, lp

Head: 3973487: 350ci, 200hp, 255hp   3998993: 350ci, 200hp, uu
3998916: 350ci, 255hp, uu   3999241: 454ci, 270hp

Carburetor: Rochester Q-jet #7042202: 350ci, 200hp, at
Rochester Q-jet #7042203: 350ci, 200hp, mt
Rochester Q-jet #7042216: 454ci, 270hp, at
Rochester Q-jet #7042217: 454ci, 270hp, mt
Rochester Q-jet #7042902: 350ci, 200hp, at, ec
Rochester Q-jet #7042903: 350ci, 200hp, mt, ec
Holley R6239A #3999263: 350ci, 255hp

Distributor: 1112050: 350ci, 200hp   1112101: 350ci, 255hp
1112051: 454ci, 270hp

Alternator: 1100543: 454ci, 270hp   1100544: All with ac
1100950: 350ci, 200hp, 255hp

Ending Vehicle:
Aug 71: 501344	Dec 71: 510310	Apr 72: 519993
Sep 71: 503697	Jan 72: 512661	May 72: 522611
Oct 71: 506050	Feb 72: 515020	Jun 72: 525226
Nov 71: 508406	Mar 72: 517613	Jul 72: 527004

Abbreviations: ac=air conditioning, ar=air injection reactor, at=automatic transmission, ci=cubic inch, ec=emission control, hp=horsepower, lp=late production, mt=manual transmission, uu=uncertain usage.

1972 Facts

• Appearance changes were minimal for 1972, but this model's significance is more associated with its "end of an era" status than by new looks or features. The 1972 Corvette was the last to feature both front and rear chrome bumpers, a bright egg-crate grill, side-fender grills (later models do have functional vents and some have vent trim). Also, the 1972 model was the last to feature the removable rear window, a feature unique to 1968 through 1972 models.

• The 1972 did not have the fiber-optics light-monitoring system used in 1968-1971 models, but the previously optional alarm (sounding horn type) was included in its base price.

• The RPO ZR2 package was not available this year, but ZR1 was and it was the last year for it and the LT1 engine. The ZR1 and LT1 designations earned special enthusiast reverence, so Chevrolet resurrected both later. The ZR1 was reborn as a special 1990-1995 model with an all-aluminum engine designed by Lotus and Chevrolet. LT1 designated a new and significantly improved base engine starting in 1992.

• This was the only year in the 1970-1972 series that the LT1 engine could be combined with air conditioning. The number built is believed to be 240. Previous restrictions were based on the fear of air conditioning belts being spun off by the higher engine revolutions permitted by solid valve lifters. The only previous Corvette availability of the solid lifter and air conditioning combination was with 1964-65 models with L76 365hp engines. To discourage high rpm with 1972's application, LT1s with air conditioning had the base engine's 5600 rpm tach instead of the 6500 rpm unit used with non-air LT1s.

1972 Options

RPO #	DESCRIPTION	QTY	RETAIL $
19437	Base Corvette Sport Coupe	20,496	$5,533.00
19467	Base Corvette Convertible	6,508	5,296.00
—	Custom Interior Trim	8,709	158.00
AV3	Three Point Seat Belts	17,693	—
A31	Power Windows	9,495	85.35
A85	Custom Shoulder Belts (std with coupe)	749	42.15
C07	Auxiliary Hardtop (for convertible)	2,646	273.85
C08	Vinyl Covering (for auxiliary hardtop)	811	158.00
C50	Rear Window Defroster	2,221	42.15
C60	Air Conditioning	17,011	464.50
ZQ1	Optional Rear Axle Ratio	1,986	12.65
J50	Power Brakes	18,770	47.40
K19	Air Injection Reactor	3,912	—
LS5	454ci, 270hp Engine (n/a California)	3,913	294.90
LT1	350ci, 255hp Engine	1,741	483.45
M21	4-Speed Manual Trans, close ratio	1,638	0.00
M40	Turbo Hydra-Matic Automatic Transmission	14,543	0.00
N37	Tilt-Telescopic Steering Column	12,992	84.30
N40	Power Steering	23,794	115.90
P02	Deluxe Wheel Covers	3,593	63.20
PT7	White Stripe Tires, F70x15, nylon	6,666	30.35
PU9	White Letter Tires, F70x15, nylon	16,623	43.65
T60	Heavy Duty Battery (std with LS5)	2,969	15.80
U69	AM-FM Radio	19,480	178.00
U79	AM-FM Radio, stereo	7,189	283.35
YF5	California Emission Test	1,967	15.80
ZR1	Special Purpose LT1 Engine Package	20	1,010.05

• Repeal of Federal excise tax on 12/11/71 reduced retail prices.

• A 350ci, 200hp engine, 4-speed wide-ratio manual transmission, vinyl interior trim, and soft top (conv) or T-tops were included in the base.

• Horsepower rating decreases in 1972 were due mainly to a change in the measuring criteria. Previously, horsepower ratings were measured as the "gross" output of the engine alone. Starting in 1972, ratings were a more realistic, real-world "net" which included losses from air cleaners, exhaust systems and accessories.

• The ZR1 package included the LT1 engine, heavy-duty close-ratio 4-speed manual transmission, heavy-duty power brakes, transistor ignition, special aluminum radiator, and special springs, shocks, and front and rear stabilizer bars (ZR1s have appeared with and without rear stabilizers). 1972 ZR1s generally did not have fan shrouds. RPOs A31, C50, C60, N40, P02, U69 and U79 were not available with ZR1.

• M40 was no cost with the base 350ci, 200-hp engine, but cost $97 with LS5 (454ci, 270hp). It was not available with LT1 or ZR1.

1972 Colors

CODE	EXTERIOR	QTY	SOFT TOP	INTERIORS
912	Sunflower Yellow	1,543	Bk-W	Bk-S
924	Pewter Silver	1,372	Bk-W	B-Bk-R-S
945	Bryar Blue	1,617	Bk-W	Bk
946	Elkhart Green	4,200	Bk-W	Bk-S
972	Classic White	2,763	Bk-W	B-Bk-R-S
973	Mille Miglia Red	2,478	Bk-W	Bk-R-S
979	Targa Blue	3,198	Bk-W	B-Bk
987	Ontario Orange	4,891	Bk-W	Bk-S
988	Steel Cities Gray	2,346	Bk-W	Bk-R-S
989	War Bonnet Yellow	2,550	Bk-W	Bk-S

• Suggested interiors shown. Additional combinations were possible.

• Paint quantities do not add to total production because additional units had non-standard paint, including special colors and primer paint override. All 1972 wheels were painted silver.

• Seat and shoulder belts matched interior colors except for the blue interior which received darker blue belts.

Interior Codes: 400=Bk/V, 404=Bk/L, 407=R/V, 412=B/V, 417=S/V, 421=S/L.

Abbreviations: B=Blue, Bk=Black, L=Leather, R=Red, S=Saddle, V=Vinyl, W=White.

1973 Corvette

Production: 25,521 coupe, 4,943 convertible, 30,464 total.

1973 Numbers

Vehicle: 1Z37J3S400001 through 1Z37J3S434464
- For convertibles, third digit is a 6.
- 4,000 vehicle identification numbers were not used.
- Fifth digit varies as follows: J=350ci, 190hp; T=350ci, 250hp; Z=454ci, 275hp.

Suffix:

CKZ: 350ci, 190hp, mt	CLR: 350ci, 250hp, mt
CLA: 350ci, 190hp, at	CLS: 350ci, 250hp, mt, ce
CLB: 350ci, 190hp, mt, ce	CWM: 454ci, 275hp, mt
CLC: 350ci, 190hp, at, ce	CWR: 454ci, 275hp, at
CLD: 350ci, 250hp, at	CWS: 454ci, 275hp, at, ce
CLH: 350ci, 250hp, at, ce	CWT: 454ci, 275hp, mt, ce

Block: 3970010, 3970014: 350ci, 190hp, 250hp   3999289: 454ci, 275hp

Head: 353049: 454ci, 275hp   330545: 350ci, 250hp
3998993, 333881, 333882: 350ci, 190hp

Carburetor: Rochester Q-jet #7043200: 454ci, 275hp, at
Rochester Q-jet #7043201: 454ci, 275hp, mt
Rochester Q-jet #7043202: 350ci, 190hp, at
Rochester Q-jet #7043203: 350ci, 190hp, mt
Rochester Q-jet #7043212: 350ci, 250hp, at
Rochester Q-jet #7043213: 350ci, 250hp, mt

Distributor: 1112098: 350ci, 190hp   1112130: 350ci, 250hp, mt
1112114: 454ci, 275hp   1112150: 350ci, 250hp, at

Alternator: 1100544: 350ci, 454ci, ac   1102353: 454ci
1100950: 350ci, 454ci

Ending Vehicle:

Aug 72: 01138	Dec 72: 10679	Apr 73: 21933
Sep 72: 03539	Jan 73: 13600	May 73: 28892
Oct 72: 06054	Feb 73: 16301	Jun 73: 31731
Nov 72: 08696	Mar 73: 19253	Jul 73: 34464

Abbreviations: ac=air conditioning, at=automatic transmission, ce=california emissions, ci=cubic inch, hp=horsepower, mt=manual transmission.

1973 Facts

• In the 1973 model year, 4,000 serial numbers were never built. The last 1973 Corvette's serial number ends with 34,464, but production totaled 30,464. The unused VIN numbers were 24,001 through 28,000.

• Rear 1973 bumpers were essentially unchanged from 1972, but fronts were redesigned to meet federal 5-mph standards. The front bumper system, with body-color, injected molded urethane cover, added thirty-five pounds of weight and could withstand 5-mph impacts without damage to lights or safety equipment.

• A new coolant recovery system routed high temperature overflow coolant into a holding reservoir for return to the radiator after cooling.

• Material was sprayed on several inner panels for sound deadening, and a new sound-absorbing pad was installed on the inner-hood surface. Chevrolet ads claimed a forty-percent reduction in cabin noise levels.

• New chassis mounts were used in 1973 to better isolate road chatter and vibration. The mounts were rubber with steel sleeves. The rubber provided vertical cushioning while the steel maintained stability.

• The lifting windshield wiper panel was deleted from 1973 models, but a new hood with rear cold air induction was introduced.

• Steel beams were added inside doors to improve side impact protection.

• Radial tires were first used with Corvettes in 1973.

• Rear windows weren't removable in 1973, but rear luggage space height increased two inches due to removal of the window storage shelf. Speaking to *Car and Driver* magazine, Zora Arkus-Duntov stated that removal of the rear window created too much interior air buffeting and that this, not cost savings, was the reason for a fixed window in 1973.

1973 Options

RPO #	DESCRIPTION	QTY	RETAIL $
1YZ37	Base Corvette Sport Coupe	25,521	$5,561.50
1YZ67	Base Corvette Convertible	4,943	5,398.50
—	Custom Interior Trim	13,434	154.00
A31	Power Windows	14,024	83.00
A85	Custom Shoulder Belts (std with coupe)	788	41.00
C07	Auxiliary Hardtop (for convertible)	1,328	267.00
C08	Vinyl Covering (for auxiliary hardtop)	323	62.00
C50	Rear Window Defroster	4,412	41.00
C60	Air Conditioning	21,578	452.00
—	Optional Rear Axle Ratio	1,791	12.00
J50	Power Brakes	24,168	46.00
LS4	454ci, 275hp Engine	4,412	250.00
L82	350ci, 250hp Engine	5,710	299.00
M21	4-Speed Manual Trans, close ratio	3,704	0.00
M40	Turbo Hydra-Matic Automatic Transmission	17,927	0.00
N37	Tilt-Telescopic Steering Column	17,949	82.00
N40	Power Steering	27,872	113.00
P02	Deluxe Wheel Covers	1,739	62.00
QRM	White Stripe Steel Belted Tires, GR70x15	19,903	32.00
QRZ	White Letter Steel Belted Tires, GR70x15	4,541	45.00
T60	Heavy Duty Battery (standard with LS4)	4,912	15.00
U58	AM-FM Radio, stereo	12,482	276.00
U69	AM-FM Radio	17,598	173.00
UF1	Map Light (on rearview mirror)	8,186	5.00
YF5	California Emission Test	3,008	15.00
YJ8	Cast Aluminum Wheels (5)	4	175.00
Z07	Off Road Suspension and Brake Package	45	369.00

• A 350ci, 190hp engine, 4-speed wide-ratio manual transmission, soft top (convertible) or T-tops (coupe), and vinyl interior trim were included in the base price.

• The Z07 package was available only with L82 or LS4 engines, and required J50 and M21. It included special front and rear suspension and heavy-duty power front and rear brakes. It was not available with C60.

• YJ8 cast aluminum wheels were offered, but sales records show only four sets (five per set) sold. Early in production, Chevrolet rejected the wheels for porosity problems and recalled the wheels that had been released. Rumors place the number of wheels produced at 800 sets. Some wheels are definitely in private hands, but the actual number remaining is unknown. The casting number on these rare wheels was 329381. Lug nuts had a recessed center area painted black and were unique to this Corvette application.

• Custom interior included leather seat trim, wood-grain accents and lower carpet trim on interior door panels, wood-grain accents on console, and special cut-pile carpeting.

• M40 was no cost with base engine, but cost $97 with LS4 or L82.

1973 Colors

CODE	EXTERIOR	SOFT TOP	WHEELS	INTERIORS
910	Classic White	Bk-W	Silver	Bk-Dr-Ds-Mb-Ms
914	Silver	Bk-W	Silver	Bk-Dr-Ds-Mb-Ms
922	Medium Blue	Bk-W	Silver	Bk-Mb-Ms
927	Dark Blue	Bk-W	Silver	Bk-Dr-Mb-Ms
945	Blue-Green	Bk-W	Silver	Bk-Dr-Ds-Ms
947	Elkhart Green	Bk-W	Silver	Bk-Ms
952	Yellow	Bk-W	Silver	Bk-Ds-Mb
953	Metallic Yellow	Bk-W	Silver	Bk-Mb
976	Mille Miglia Red	Bk-W	Silver	Bk-Dr-Ds-Mb-Ms
980	Orange	Bk-W	Silver	Bk-Ms-Mb-Ds

• Suggested interiors shown. Other combinations were possible.

• Deluxe interior, which included leather seating, was available in black, medium saddle, and dark saddle only.

• Exterior color quantities are currently not available from Chevrolet. It is believed, however, that in addition to the colors listed, about 30 cars were painted black at the St. Louis assembly plant.

Interior Codes: 400=Bk/V, 404=Bk/L, 413=Mb/V, 415=Ms/V, 416=Ms/L, 418=Ds/V, 422=Ds/L, 425=Dr/V.

Abbreviations: Bk=Black, Dr=Dark Red, Ds=Dark Saddle, L=Leather, Mb=Midnight Blue, Ms=Medium Saddle, V=Vinyl, W=White.

1974 Corvette

Production: 32,028 coupe, 5,474 convertible, 37,502 total.

1974 Numbers

Vehicle: 1Z37J4S400001 through 1Z37J4S437502
- For convertibles, third digit is a 6.
- Fifth digit varies as follows: J=350ci, 195hp; T=350ci, 250hp; Z=454ci, 270hp.

Suffix:
CKZ: 350ci, 195hp, mt
CLA: 350ci, 195hp, at
CLB: 350ci, 195hp, mt, ce
CLC: 350ci, 195hp, at, ce
CLD: 350ci, 250hp, at
CLH: 350ci, 250hp, at, ce
CLR: 350ci, 250hp, mt
CLS: 350ci, 250hp, mt, ce
CWM: 454ci, 270hp, mt
CWR: 454ci, 270hp, at
CWS: 454ci, 270hp, at, ce
CWT: 454ci, 270hp, mt, ce

Block: 3970010: 350ci, 195hp, 250hp
3999289: 454ci, 270hp

Head: 333881: 350ci, 195hp   333882: 350ci, 195hp, 250hp
336781: 454ci, 270hp

Carburetor: Rochester Q-jet #7044221: 454ci, 270hp, mt
Rochester Q-jet #7044206: 350ci, 195hp, at
Rochester Q-jet #7044207: 350ci, 195hp, mt
Rochester Q-jet #7044210: 350ci, 250hp, at
Rochester Q-jet #7044211: 350ci, 250hp, mt
Rochester Q-jet #7044225: 454ci, 270hp, at
Rochester Q-jet #7044505: 454ci, 270hp, at, ce
Rochester Q-jet #7044506: 350ci, 250hp, at, ce
Rochester Q-jet #7044507: 350ci, 250hp, mt, ce

Distributor: 1112114, 1112526: 454ci, 270hp   1112850: 350ci, 195hp, mt, ce
1112150: 350ci, 250hp   1112851: 350ci, 195hp, at, ce
1112247: 350ci, 195hp   1112853: 350ci, 250hp, at
1112544: 350ci, 195hp, mt, ce

Alternator: 1100544: 350ci, 454ci, ac   1100950: 350ci, 454ci
1102353: 454ci

Ending Vehicle:

Aug 73: 01250	Jan 74: 16184	Jun 74: 33257
Sep 73: 04111	Feb 74: 19258	Jul 74: 33257
Oct 73: 07605	Mar 74: 22367	Aug 74: 33257
Nov 73: 10813	Apr 74: 25751	Sep 74: 37251
Dec 73: 12830	May 74: 29602	Oct 74: 37502

- A labor dispute closed the St. Louis Corvette assembly plant in 1974 from about June 28, 1974 to September 2, 1974.

Abbreviations: ac=air conditioning, at=automatic transmission, ce=california emissions, ci=cubic inch, hp=horsepower, mt=manual trans.

1974 Facts

- The transition to "soft" bumpers was completed in 1974 with the new body-color rear bumpers. The urethane plastic skin had built-in recesses for the license plate and taillights, and a vertical center seam divided the two main sections (later years are one-piece without the seam). The skin covered an aluminum impact bar mounted on two telescopic brackets.
- This was the last Corvette model without catalytic converters. It was the last Corvette of the era with true dual exhausts. Resonators were added to the exhaust system. Exhaust tailpipes exited below the bumper. Fuel requirement was 91-octane leaded or low lead.
- Radiators were redesigned for more efficient cooling at low speeds.
- Shoulder belts on 1974 coupe models were integrated with the lap belts for the first time. Shoulder belts remained optional with convertibles and if so equipped were separate from the lap belt as before. Also, the belt locking mechanism was changed from a pull-rate type to a swinging-weight type activated by the car's deceleration.
- The alarm activator moved from the rear panel to the driver-side fender.
- Magnets were added to power steering units to attract fluid debris.
- The stock inside rearview mirror increased in width from 8 to 10 inches.

1974 Options

RPO #	DESCRIPTION	QTY	RETAIL $
1YZ37	Base Corvette Sport Coupe	32,028	$6,001.50
1YZ67	Base Corvette Convertible	5,474	5,765.50
—	Custom Interior Trim	19,959	154.00
A31	Power Windows	23,940	86.00
A85	Custom Shoulder Belts (std with coupe)	618	41.00
C07	Auxiliary Hardtop (for convertible)	2,612	267.00
C08	Vinyl-Covered Auxiliary Hardtop	367	329.00
C50	Rear Window Defroster	9,322	43.00
C60	Air Conditioning	29,397	467.00
FE7	Gymkhana Suspension	1,905	7.00
—	Optional Rear Axle Ratios	1,219	12.00
J50	Power Brakes	33,306	49.00
LS4	454ci, 270hp Engine	3,494	250.00
L82	350ci, 250hp Engine	6,690	299.00
M21	4-Speed Manual Trans, close ratio	3,494	0.00
M40	Turbo Hydra-Matic Automatic Transmission	25,146	0.00
N37	Tilt-Telescopic Steering Column	27,700	82.00
N41	Power Steering	35,944	117.00
QRM	White Stripe Steel Belted Tires, GR70x15	9,140	32.00
QRZ	White Letter Steel Belted Tires, GR70x15	24,102	45.00
U05	Dual Horns	5,258	4.00
U58	AM-FM Radio, stereo	19,581	276.00
U69	AM-FM Radio	17,374	173.00
UA1	Heavy Duty Battery (std with LS4)	9,169	15.00
UF1	Map Light (on rearview mirror)	16,101	5.00
YF5	California Emission Test	—	20.00
Z07	Off Road Suspension and Brake Package	47	400.00

• A 350ci, 195hp engine, 4-speed wide-ratio manual transmission, soft top (conv) or T-tops (cpe), and vinyl interior were included in the base.

• YJ8 cast aluminum wheels appeared on early 1974 option lists, but records indicate none were sold. Since P02 Deluxe Wheel Covers were no longer available, there were no optional wheels or covers for 1974.

• Custom interior included leather seat trim, wood-grain accents and lower carpet trim on interior door panels, wood-grain accents on console, and special cut-pile carpeting.

• M40 was no cost with the base 350ci, 195hp engine, but cost $103 ($97 early in production) with LS4 or L82. It was not available with Z07.

• The Z07 package was available only with L82 and LS4 engines, and required M21. It included special front and rear suspension, and heavy-duty front and rear power brakes.

• The FE7 gymkhana suspension included stiffer front sway bar and stiffer springs. It was included with Z07. There were no order restrictions.

• This was the last year for 454ci "big block" engines in Corvettes.

1974 Colors

CODE	EXTERIOR	SOFT TOP	WHEELS	INTERIORS
10	Classic White	Bk-W	Silver	Bk-Db-Dr-N-S-Si
14	Silver Mist	Bk-W	Silver	Bk-Db-Dr-S-Si
17	Corvette Gray	Bk-W	Silver	Bk-Db-Dr-N-S-Si
22	Corvette Med Blue	Bk-W	Silver	Bk-Db-Si
48	Dark Green	Bk-W	Silver	Bk-N-S-Si
56	Bright Yellow	Bk-W	Silver	Bk-N-S-Si
68	Dark Brown	Bk-W	Silver	Bk-N-S-Si
74	Medium Red	Bk-W	Silver	Bk-Dr-N-S-Si
76	Mille Miglia Red	Bk-W	Silver	Bk-Dr-N-S-Si
80	Corvette Orange	Bk-W	Silver	Bk-N-S-Si

• Suggested interiors shown. Other combinations were possible.

• Exterior color code on trim plates was generally a two number code followed by the letter L. The code may also appear with a 9 before the two numbers and no letter following.

• Custom interior included leather seating. Quantities were 7,828 black, 2,797 silver, 9,334 saddle.

Interior Codes: 400=Bk/V, 404=Bk/L, 406=Si/V, 407=Si/L, 408=N/V, 413=Db/V, 415=S/V, 416=S/L, 425=Dr/V.

Abbreviations: Bk=Black, Db=Dark Blue, Dr=Dark Red, L=Leather, N=Neutral, S=Saddle, Si=Silver, V=Vinyl W=White.

1975 Corvette

Production: 33,836 coupe, 4,629 convertible, 38,465 total.

1975 Numbers

Vehicle: 1Z37J5S400001 through 1Z37J5S438465
- For convertibles, third digit is a 6.
- Fifth digit varies as follows: J=350ci, 165hp
  T=350ci, 205hp

Suffix:

CHA: 350ci, 165hp, mt	CRK: 350ci, 165hp, at
CHB: 350ci, 165hp, at	CRL: 350ci, 205hp, mt
CHC: 350ci, 205hp, mt	CRM: 350ci, 205hp, at
CHR: 350ci, 205hp, at, ce	CUA: 350ci, 165hp, mt
CHU: 350ci, 165hp, mt	CUB: 350ci, 165hp, mt
CHZ: 350ci, 165hp, at, ce	CUD: 350ci, 205hp, mt
CKC: 350ci, 205hp, at,	CUT: 350ci, 205hp, mt
CRJ: 350ci, 165hp, mt	

Block: 3970010: 350ci, 165hp, 205hp

Head: 333882: 350ci, 165hp, 205hp

Carburetor: Rochester Q-jet #7045210: 350ci, 205hp, at
Rochester Q-jet #7045211: 350ci, 205hp, mt
Rochester Q-jet #7045222: 350ci, 165hp, at
Rochester Q-jet #7045223: 350ci, 165hp, mt

Distributor: 1112880: 350ci, 165hp, ce   1112888: 350ci, 165hp
1112883: 350ci, 205hp, fd   1112979: 350ci, 205hp, sd

Alternator: 1100544: With ac   1100597: hdb, fd
1102474: With ac, lp   1102487: hdb, sd
1100950: Without ac   1102484: Without ac, lp

Ending Vehicle:

Oct 74: 02385	Feb 75: 17112	Jun 75: 33474
Nov 74: 06180	Mar 75: 20856	Jul 75: 38465
Dec 74: 09190	Apr 75: 25228	
Jan 75: 13159	May 75: 29379	

Abbreviations: ac=air conditioning, at=automatic transmission, ce=california emissions, ci=cubic inch, fd=first design, hdb=heavy-duty battery, hp=horsepower, lp=late production, mt=manual transmission, sd=second design.

1975 Facts

• The 1975 convertible was for a time the "last," but convertibles returned in 1986. The last 1975 convertible was built in late July, 1975. This was the last Corvette convertible to cost less than its coupe counterpart.

• Catalytic converters first appeared in Corvettes with the 1975 model. Dual exhausts were routed to a single converter, then split for dual exit.

• Soft bumpers were redesigned structurally for 1975, and external appearance varied slightly from the previous year. The front bumper had an inner honeycomb core and simulated "pads" on the exterior. The rear bumper had inner shock absorbers for impact. The rear bumper skin was one-piece, unlike the two-piece 1974 unit. Simulated pads like those of the front were also added to the rear.

• High Energy Ignition (HEI) appeared first in Corvettes on 1975 models. Quite different from the transistor ignitions previously available, HEI included the Corvette's first no-points distributor. This also ended Corvette's distributor-driven tachometers. Starting in 1975, tachs were electronically driven.

• Optional engine choice for 1975 was limited to one, the L82. Not since 1955 was the Corvette customer's choice so restricted. And it was the first year since 1964 that only one engine displacement was offered.

• Hood emblems designating the L82 engine first appeared in 1975, but some 1975 L82 models were built without the emblems.

• Kilometer-per-hour subfaces appeared on Corvette's speedometers first in 1975; it was also the first year with a headlight warning buzzer. It was the last year for Astro Ventilation.

• This was the first year to use an internal bladder in the fuel tank.

1975 Options

RPO #	DESCRIPTION	QTY	RETAIL $
1YZ37	Base Corvette Sport Coupe	33,836	$6,810.10
1YZ67	Base Corvette Convertible	4,629	6,550.10
—	Custom Interior Trim	—	154.00
A31	Power Windows	28,745	93.00
A85	Custom Shoulder Belts (std with coupe)	646	41.00
C07	Auxiliary Hardtop (for convertible)	2,407	267.00
C08	Vinyl Covered Auxiliary Hardtop (conv)	279	350.00
C50	Rear Window Defrester	13,760	46.00
C60	Air Conditioning	31,914	490.00
FE7	Gymkhana Suspension	3,194	7.00
—	Optional Rear Axle Ratios	1,969	12.00
J50	Power Brakes	35,842	50.00
L82	350ci, 205hp Engine	2,372	336.00
M21	4-Speed Manual Trans, close ratio	1,057	0.00
M40	Turbo Hydra-Matic Automatic Transmission	28,473	0.00
N37	Tilt-Telescopic Steering Column	31,830	82.00
N41	Power Steering	37,591	129.00
QRM	White Stripe Steel Belted Tires, GR70x15	5,233	35.00
QRZ	White Letter Steel Belted Tires, GR70x15	30,407	48.00
U05	Dual Horns	22,011	4.00
U58	AM-FM Radio, stereo	24,701	284.00
U69	AM-FM Radio	12,902	178.00
UA1	Heavy Duty Battery	16,778	15.00
UF1	Map Light (on rearview mirror)	21,676	5.00
YF5	California Emission Test	3,037	20.00
Z07	Off Road Suspension and Brake Package	144	400.00

• A 350ci, 165hp engine, 4-speed wide-ratio manual transmission, soft top (convertible) or T-tops (coupe), and vinyl interior were included in the base price.

• Custom interior included leather seat trim, wood-grain accents and lower carpet trim on inner door panels, wood-grain accents on console, and special cut-pile carpeting.

• The Z07 package was available only with L82 engines and required M21 transmission. It included FE7's special front and rear suspension, plus heavy-duty front and rear power brakes..

• The FE7 gymkhana suspension included stiffer front stabilizer bar and stiffer springs. It was included with Z07. There were no engine or transmission order restrictions with FE7.

• M40 was no cost with the base 350ci, 165hp engine, but cost $120.00 with optional L82 engine. M21 was no cost but required optional L82.

1975 Colors

CODE	EXTERIOR	QTY	SOFT TOP	INTERIORS
10	Classic White	8,007	Bk-W	Bk-Db-Dr-Ms-N-Si
13	Silver	4,710	Bk-W	Bk-Db-Dr-Ms-Si
22	Bright Blue	2,869	Bk-W	Bk-Db-Si
27	Steel Blue	1,268	Bk-W	Bk-Db-Si
42	Bright Green	1,664	Bk-W	Bk-Ms-N-Si
56	Bright Yellow	2,883	Bk-W	Bk-Ms-N
67	Medium Saddle	3,403	Bk-W	Bk-Ms-N
70	Orange Flame	3,030	Bk-W	Bk-Ms-N
74	Dark Red	3,342	Bk-W	Bk-Dr-Ms-N-Si
76	Mille Miglia Red	3,355	Bk-W	Bk-Dr-Ms-N-Si

• Suggested interiors shown. Additional combinations were possible.

• Paint quantities do not add to total production, possibly because additional units had non-standard paint, or primer only.

• Custom Interior Trim, which included leather seating, was available in black, silver, dark blue, medium saddle, and dark red.

• All 1975 wheels were painted silver.

• Steel Blue exterior was available for approximately three months.

• Silver vinyl interior quantity sold in 1975 was 992.

Interior Codes: 19V=Bk/V, 192=Bk/L, 14V=Si/V, 142=Si/L, 26V=Db/V, 262=Db/L, 60V=N/V, 65V=Ms/V, 652=Ms/L, 73V=Dr/V, 732=Dr/L.

Abbreviations: Bk=Black, Db=Dark Blue, Dr=Dark Red, L=Leather, Ms=Medium Saddle, N=Neutral, Si=Silver, V=Vinyl, W=White.

1976 Corvette
Production: 46,558 coupes.

1976 Numbers

Vehicle: 1Z37L6S400001 through 1Z37L6S446558
- Fifth digit varies as follows: L=350ci, 180hp
  X=350ci, 210hp

Suffix: CHC: 350ci, 210hp, mt   CKX: 350ci, 180hp, at
CKC: 350ci, 210hp, at   CLS: 350ci, 180hp, at, ce
CKW: 350ci, 180hp, mt

Block: 3970010: 350ci, 180hp, 210hp

Head: 333882: 350ci, 180hp, 210hp

Carburetor: Rochester Q-jet #17056206: 350ci, 180hp, at
Rochester Q-jet #17056207: 350ci, 180hp, mt
Rochester Q-jet #17056210: 350ci, 210hp, at
Rochester Q-jet #17056211: 350ci, 210hp, mt
Rochester Q-jet #17056226: 350ci, 210hp, at, ac
Rochester Q-jet #17056506: 350ci, 180hp, at, ce
Rochester Q-jet #17056507: 350ci, 180hp, mt, ce

Distributor: 1103200: 350ci, 210hp,mt   1112905: 350ci, 180hp, at, ce
1112888: 350ci, 180hp   1112979: 350ci, 210hp, at

Alternator: 1102474: All with ac   1102484: All without ac

Ending Vehicle: Aug 75: 01602   Jan 76: 20568   Jun 76: 40830
Sep 75: 05693   Feb 76: 24370   Jul 76: 44767
Oct 75: 09982   Mar 76: 28760   Aug 76: 46558
Nov 75: 13481   Apr 76: 32805
Dec 75: 16696   May 76: 36656

Abbreviations: ac=air conditioning, at=automatic transmission, ce=california emissions, ci=cubic inch, hp=horsepower, mt=manual transmission.

1976 Facts

- The carburetor air induction system was revised in 1976. Previously, air was drawn in at the rear of the hood, producing a slight howl audible from within the car. The air source point was moved forward, so that air was pulled in over the radiator. The 1976 hood is unique to the year.

- The aluminum wheels announced for 1973 arrived as a bona fide option with the 1976 model. These were made by Kelsey Hayes in Mexico and the wheels are identified on their inside surfaces as to source and build location. The YJ8 option included four wheels and a conventional steel spare, in contrast to the five-wheel sets planned for 1973, and to the five-wheel knock-off and bolt-on aluminum wheel options of 1963-1967.

- Engineers put a partial steel underbelly in the forward section of Corvettes starting in 1976 for added rigidity, weight reduction, and to better isolate the cockpit from heat generated by catalytic converters. Also, higher engine temperatures were intentional, one way to increase efficiency and partially offset emissions-related power losses.

- There is documented evidence that the final VIN for 1976 ended in 446567, nine more units than the published total of 46,558.

- Astro Ventilation was eliminated starting with 1976 models. This meant, among other things, that vents on the rear deck (just aft of the back window) were deleted.

- A new "sport" steering wheel for 1976 Corvettes was shared by Chevrolet Vegas (and other Chevrolet models), a development not viewed well by Corvette enthusiasts.

- GM's "freedom" battery, a new sealed and maintenance-free unit, was included with all 1976 models.

- Some late-production 1976 Corvettes were built with parts normally associated with 1977 models, especially interior components.

- Two styles of rear bumpers were used. The first had smaller, recessed "Corvette" letters. The second style had larger letters, not recessed.

- The 1976 was the last of the era to require a unique Delco radio due to the available console depth.

1976 Options

RPO #	DESCRIPTION	QTY	RETAIL $
1YZ37	Base Corvette Sport Coupe	46,558	$7,604.85
—	Custom Interior Trim	36,762	164.00
A31	Power Windows	38,700	107.00
C49	Rear Window Defogger	24,960	78.00
C60	Air Conditioning	40,787	523.00
FE7	Gymkhana Suspension	5,368	35.00
—	Optional Rear Axle Ratios	1,371	13.00
J50	Power Brakes	46,558	59.00
L82	350ci, 210hp Engine	5,720	481.00
M21	4-Speed Manual Trans, close ratio	2,088	0.00
M40	Turbo Hydra-Matic Automatic Transmission	36,625	0.00
N37	Tilt-Telescopic Steering Column	41,797	95.00
N41	Power Steering	46,385	151.00
QRM	White Stripe Steel Belted Tires, GR70x15	3,992	37.00
QRZ	White Letter Steel Belted Tires, GR70x15	39,923	51.00
U58	AM-FM Radio, stereo	34,272	281.00
U69	AM-FM Radio	11,083	187.00
UA1	Heavy Duty Battery	25,909	16.00
UF1	Map Light (on rearview mirror)	35,361	10.00
YF5	California Emission Test	3,527	50.00
YJ8	Aluminum Wheels (4)	6,253	299.00

• A 350ci, 180hp engine, 4-speed wide-ratio manual transmission, T-Tops, and vinyl interior trim were included in the base price.

• Custom interior included leather seat trim, wood-grain accents and lower carpet trim on inner door panels, wood grain accents on console, and special cut-pile carpeting.

• The FE7 gymkhana suspension included stiffer front stabilizer bar and stiffer springs. FE7 could be ordered with any engine or transmission.

• M40 was no cost with the base 350ci, 180hp engine, but cost $134.00 with optional L82 engine. M21 was no cost but required optional L82.

• The only engine-transmission combination available in California was the base 350ci, 180hp engine with M40 automatic transmission.

• Listed as separate options initially, power brakes (J50) and power steering (N41) were included in an increased base price during 1976. All 1976 Corvettes had power brakes; all but 173 had power steering.

• RPO C49 used glass heating elements instead of forced air and the terminology changed from "defroster" to "defogger."

• In 1976, 1,203 Corvettes were ordered without a radio.

• Sales to Canada for 1976 were 4,289.

1976 Colors

CODE	EXTERIOR	QTY	WHEELS	INTERIORS
10	Classic White	10,674	Silver	Bk-Bg-Bu-Db-F-Sg-W
13	Silver	6,934	Silver	Bk-Bg-Bu-F-Sg-W
22	Bright Blue	3,268	Silver	Bk-Sg
33	Dark Green	2,038	Silver	Bk-Bg-Bu-Sg-W
37	Mahogany	4,182	Silver	Bk-Bu-F-Sg-W
56	Bright Yellow	3,389	Silver	Bk-Db
64	Buckskin	2,954	Silver	Bk-Bu-Db-F-W
69	Dark Brown	4,447	Silver	Bk-Bu-Db-W
70	Orange Flame	4,073	Silver	Bk-Bu-Db
72	Red	4,590	Silver	Bk-Bu-F-Sg-W

• Suggested interiors shown. Additional combinations were possible.

• Paint quantities do not add to total production because additional units had nonstandard paint, and primer only.

• Early Chevrolet order guides show an exterior code 39 for Dark Green Metallic. This code was changed to code 33, but production records indicate one code 39 Dark Green Metallic 1976 Corvette built.

• Known interior color quantities for 1976 are 3,258 black vinyl, 7,826 black leather, 3,256 buckskin vinyl, 8,946 buckskin leather, 6,184 dark brown leather, 3,281 firethorn vinyl, 8,259 firethorn leather.

Interior Codes: 112=W/L, 15V=W/V, 152=Sg/L, 19V=Bk/V, 192=Bk/L, 322=Bg/L, 64V=Bu/V, 642=Bu/L, 692=Db/L, 71V=F/V, 712=F/L.

Abbreviations: Bg=Blue-Green, Bk=Black, Bu=Buckskin, Db=Dark Brown, F=Firethorn, L=Leather, Sg=Smoked Grey, V=Vinyl, W=White.

1977 Corvette

Production: 49,213 coupes.

1977 Numbers

Vehicle: 1Z37L7S400001 through 1Z37L7S449213
- Fifth digit varies as follows: L=350ci, 180hp
  X=350ci, 210hp

Suffix:
CHD: 350ci, 180hp, at, ce	CLB: 350ci, 180hp, at, ha
CKD: 350ci, 180hp, at, ha	CLC: 350ci, 180hp, at, ce
CKZ: 350ci, 180hp, mt	CLD: 350ci, 210hp, mt
CLA: 350ci, 180hp, at	CLF: 350ci, 210hp, at

Block: 3970010: 350ci, 180hp, 210hp

Head: 333882: 350ci, 180hp, 210hp   374650: 350ci, 180hp

Carburetor: Rochester Q-jet #17057202: 350ci, 180hp, at
Rochester Q-jet #17057203: 350ci, 180hp, mt
Rochester Q-jet #17057204: 350ci, 180hp, at, ac
Rochester Q-jet #17057210: 350ci, 210hp, at
Rochester Q-jet #17057211: 350ci, 210hp, mt
Rochester Q-jet #17057228: 350ci, 210hp, at, ac
Rochester Q-jet #17057502: 350ci, 180hp, at, ce
Rochester Q-jet #17057504: 350ci, 180hp, at, ac, ce
Rochester Q-jet #17057510: 350ci, 210hp, at, uu, ce
Rochester Q-jet #17057582: 350ci, 180hp, at, ha
Rochester Q-jet #17057584: 350ci, 180hp, at, ac, ha

Distributor: 1103246: 350ci, 180hp   1103256: 350ci, 210hp
1103248: 350ci, 180hp, at, ce

Alternator: 1102474: ac or rd, ep   1102908: ac or rd
1102484: All without ac   1102909: ac or rd

Ending Vehicle:
Aug 76: 02287	Jan 77: 21118	May 77: 37029
Sep 76: 06337	Feb 77: 24662	Jun 77: 41233
Nov 76: 14216	Mar 77: 29041	Jul 77: 45179
Dec 76: 17551	Apr 77: 33057	Aug 77: 49213

Abbreviations: ac=air conditioning, at=automatic transmission, ce=california emissions, ci=cubic inch, ep=early production, ha=high altitude, hp=horsepower, mt=manual transmission, rd=rear defogger, uu=uncertain usage.

1977 Facts

• A new console held heater and air conditioning controls and accepted standard Delco radios due to its increased depth. A new steering column positioned the steering wheel two inches closer to the instrument panel to provide more of an "arms out" driving position, and easier entry and exit. 1977 models without RPO N37 had 1976 style steering wheels.

• The V54 roof rack was designed to hold the T-top panels, permitting use of the full luggage compartment when panels were removed.

• Early 1977 option listings contained CC1 glass roof panels ($200), but these were never available during 1977 due to a marketing exclusivity dispute between Chevrolet and the panel vendor. Chevrolet released its own glass panels in 1978 ($349); the vendor sold their panels in the aftermarket under the trade name "Moon Roofs."

• Effective with #1Z37X7S427373, the alarm activator was moved from the driver-side fender to the driver-side door lock.

• New option K30 speed control required automatic transmission.

• Leather seats were standard for the first time in 1977, but a cloth-leather combination could be substituted at no cost.

• Between August 23 and September 7, 1976, Flint Engine changed from orange paint to blue for Corvette engines. Thus, early production 1977 models were orange, later production were blue.

• The headlight dimmer and windshield wiper/washer controls were located on steering column stalks. Sunshades were redesigned to permit swinging to the side windows. The interior rearview mirror mounting point was relocated from above the windshield to the windshield itself.

1977 Options

RPO #	DESCRIPTION	QTY	RETAIL $
1YZ37	Base Corvette Sport Coupe	49,213	$8,647.65
A31	Power Windows	44,341	116.00
B32	Color Keyed Floor Mats	36,763	22.00
C49	Rear Window Defogger	30,411	84.00
C60	Air Conditioning	45,249	553.00
D35	Sport Mirrors	20,206	36.00
FE7	Gymkhana Suspension	7,269	38.00
G95	Optional Rear Axle Ratios	972	14.00
K30	Speed Control	29,161	88.00
L82	350ci, 210hp Engine	6,148	495.00
M21	4-Speed Manual Trans, close ratio	2,060	0.00
M40	Turbo Hydra-Matic Automatic Transmission	41,231	0.00
NA6	High Altitude Emission Equipment	854	22.00
N37	Tilt-Telescopic Steering Column	46,487	165.00
QRZ	White Letter Steel Belted Tires, GR70x15	46,227	57.00
UA1	Heavy Duty Battery	32,882	17.00
U58	AM-FM Radio, stereo	18,483	281.00
U69	AM-FM Radio	4,700	187.00
UM2	AM-FM Radio, stereo with 8-track tape	24,603	414.00
V54	Luggage and Roof Panel Rack	16,860	73.00
YF5	California Emission Certification	4,084	70.00
YJ8	Aluminum Wheels (4)	12,646	321.00
ZN1	Trailer Package	289	83.00
ZX2	Convenience Group	40,872	22.00

• A 350ci, 180hp engine, 4-speed wide-ratio manual transmission, T-tops, and leather interior trim were included in the base price.

• RPO ZX2 convenience group included dome light delay, headlight warning buzzer, underhood light, low fuel warning light, interior courtesy lights and right side visor mirror. Early in production, the low-fuel warning light wasn't available, so Chevrolet deleted it and reduced the cost of the option to $18. Quantity of low-fuel light deletes was 3,881.

• RPO FE7 suspension included stiffer front sway bar and stiffer springs. There were no engine or transmission order restrictions with FE7.

• RPO M40 was no cost with the base 350ci, 180hp engine, but cost $146 with optional L82 engine. M21 was no cost but required optional L82.

• The only engine-transmission combination available in California was the base 350ci, 180hp engine with M40 automatic transmission.

• RPO NA6 high altitude emission equipment was required for +4000ft; available only with the base 350ci, 180hp engine and M40 transmision.

• In 1977, 1,248 Corvettes were ordered without a radio.

• Sales to Canada for 1977 were 4,120.

1977 Colors

CODE	EXTERIOR	QTY	WHEELS	INTERIORS
10	Classic White	9,408	Silver	B-Bk-Br-Bu-R-Sg-W
13	Silver	5,518	Silver	B-Bk-R-Sg-W
19	Black	6,070	Silver	Bk-Bu-R-Sg-W
26	Corvette Light Blue	5,967	Silver	Bk-Sg-W
28	Corvette Dark Blue	4,065	Silver	B-Bk-Bu-Sg-W
41	Corvette Chartreuse	1	Silver	Bk
52	Corvette Yellow	71	Silver	Bk-Br
56	Corvette Bright Yellow	1,942	Silver	Bk-Br
66	Corvette Orange	4,012	Silver	Bk-Br-Bu
80	Corvette Tan	4,588	Silver	Bk-Br-Bu-R-W
72	Medium Red	4,057	Silver	Bk-Bu-R-Sg-W
83	Corvette Dark Red	3,434	Silver	Bk-Bu-Sg

• Suggested interiors shown. Additional combinations were possible.

• Paint quantities do not add to total production because additional units had non-standard paint, or primer only.

Interior Codes: 112=W/L, 15C=Sg/C, 152=Sg/L, 19C-Bk/C, 192=Bk/L, 27C=B/C, 272=B/L, 64C=Bu/C, 642=Bu/L, 69C=Br/C, 692=Br/L, 72C=R/C, 722=R/L.

• Leather seating was considered standard. Cloth codes designate combinations of cloth and leather which were optional at no cost.

Abbreviations: B=Blue, Bk=Black, Bu=Buckskin, Br=Brown, C=Cloth, L=Leather, R=Red, Sg=Smoked Grey, W=White.

1978 Corvette

Production: 40,274 coupe, 6,502 pace car coupe, 46,776 total

1978 Numbers

Vehicle: 1Z87L8S400001 through 1Z87L8S440274
1Z87L8S900001 through 1Z87L8S906502 (pace car)
• Fifth digit varies as follows: L=350ci, 175hp,185hp
4=350ci, 220hp

Suffix: CHW: 350ci, 185hp, mt   CMR: 350ci, 220hp, mt
CLM: 350ci, 185hp, at   CMS: 350ci, 220hp, at
CLR: 350ci, 175hp, ce, at   CUT: 350ci, 185hp, at
CLS: 350ci, 175hp, ha, at

Block: 3970010: All   376450, 460703: uncertain usage

Head: 462624: All

Carburetor: Rochester Q-jet #17058202: 350ci, 185hp, at
Rochester Q-jet #17058203: 350ci, 185hp, mt
Rochester Q-jet #17058204: 350ci, 185hp, at, ac, fd
Rochester Q-jet #17058206: 350ci, 185hp, at, ac, sd
Rochester Q-jet #17058210: 350ci, 220hp, at
Rochester Q-jet #17058211: 350ci, 220hp, mt
Rochester Q-jet #17058228: 350ci, 220hp, at, ac
Rochester Q-jet #17058502: 350ci, 175hp, at, ce
Rochester Q-jet #17058504: 350ci, 175hp, at, ac, ce
Rochester Q-jet #17058582: 350ci, 175hp, at, ha
Rochester Q-jet #17058584: 350ci, 175hp, at, ac, ha

Distributor: 1103285: 350ci, 175hp, ce   1103337: 350ci, 185hp, mt
1103291: 350ci, 220hp   1103353: 350ci, 185hp, at

Alternator: 1102474: 350ci, ac or rd, ep   1102908: 350ci, ac or rd, lp
1102484: 350ci

Abbreviations: at=automatic transmission, ce=california emissions, ci=cubic inch, ep=early production, fd=first design, ha=high altitude, hp=horsepower, lp=late production, mt=manual transmission, rd=rear defogger, sd=second design.

1978 Facts

• Chevrolet marked the Corvette's twenty-fifth year by introducing the most extensively redesigned model since the 1968. New "fastback' rear end styling featured a large rear window, but not a hatchback. This design change created significantly more luggage space behind the seats. Also, a retracting cover was added for security and sun protection.

• The 1978 interior was redesigned significantly. The speedometer and tachometer were redone in a more square, vertical mode. A glove box was added. Inner door panels were completely new and featured screwed-on armrests instead of the molded-in style common to Corvettes since 1965. Windshield wiper and washer controls were moved back to the instrument panel, but the dimmer remained on the steering column. The center console, redone for 1977, essentially carried over.

• "25th Anniversary" emblems appeared exclusively on 1978 models.

• Wider 60-series tires became available as a 1978 Corvette option and required fender trimming at the Corvette assembly plant for clearance.

• Corvette paced the 1978 Indy 500 race. To commemorate the event, Chevrolet built limited edition Corvette pace car replicas. These vehicles had their own vehicle identification number sequence, a Corvette first. Initially, these were to have two-tone silver paint with red striping, special Goodyear tires with "Corvette" sidewall letters, and a build quantity of 300 to honor the 1953 model. The special tires were eliminated and quantity ratcheted up several times as dealer and public interest skyrocketed. Chevrolet finally decided to build one for each dealer (about 6000) plus some extras. The final build quantity released by Chevrolet was 6,502; however, other quantities have been published.

• The original Pace Car paint scheme evolved into a "Silver Anniversary" paint option consisting of two-tone silver (lighter silver upper surface and darker silver lower surface), divided by silver striping. Sport mirrors and aluminum wheels were required. Quantities were not limited.

1978 Options

RPO#	DESCRIPTION	QTY	RETAIL $
1YZ87	Base Corvette Sport Coupe	40,274	$9,351.89
1YZ87/78	Limited Edition Corvette (pace car)	6,502	13,653.21
A31	Power Windows	36,931	130.00
AU3	Power Door Locks	12,187	120.00
B2Z	Silver Anniversary Paint	15,283	399.00
CC1	Removable Glass Roof Panels	972	349.00
C49	Rear Window Defogger	30,912	95.00
C60	Air Conditioning	37,638	605.00
D35	Sport Mirrors	38,405	40.00
FE7	Gymkhana Suspension	12,590	41.00
G95	Optional Rear Axle Ratio	382	15.00
K30	Cruise Control	31,608	99.00
L82	350ci, 220hp Engine	12,739	525.00
M21	4-Speed Manual Trans, close ratio	3,385	0.00
MX1	Automatic Transmission	38,614	0.00
NA6	High Altitude Emission Equipment	260	33..00
N37	Tilt-Telescopic Steering Column	37,858	175.00
QBS	White Letter SBR Tires, P255/60R15	18,296	216.32
QGR	White Letter SBR Tires, P225/70R15	26,203	51.00
UA1	Heavy Duty Battery	28,243	18.00
UM2	AM-FM Radio, stereo with 8-track tape	20,899	419.00
UP6	AM-FM Radio, stereo with CB	7,138	638.00
U58	AM-FM Radio, stereo	10,189	286.00
U69	AM-FM Radio	2,057	199.00
U75	Power Antenna	23,069	49.00
U81	Dual Rear Speakers	12,340	49.00
YF5	California Emission Certification	3,405	75.00
YJ8	Aluminum Wheels (4)	28,008	340.00
ZN1	Trailer Package	972	89.00
ZX2	Convenience Group	37,222	84.00

• A 350ci, 185hp engine, 4-speed wide-ratio manual transmission, T-tops, and leather interior trim were included in the base price.

• ZX2 included dome light delay, headlight warning buzzer, underhood light, low-fuel warning light, interior courtesy lights, floor mats, intermittent wipers (usually), and right side visor mirror.

• Pace car replica had A31, AU3, CC1, C49, C60, D35, N37, QBS, UA1, UM2, U75, U81, YJ8 (red accent), ZX2, and special seats. Other options available at normal prices except UP6 substitution for UM2 at $170.00.

• Manual transmission and/or L82 not available with NA6 or YF5.

• In 1978, 1,671 Corvettes were ordered without a radio.

• Sales to Canada for 1978 were 4,515.

1978 Colors

CODE	EXTERIOR	QTY	WHEELS	INTERIORS
10	Classic White	4,150	Silver	Bk-Db-Dbr-Lb-O-M-R
13	Silver	3,232	Silver	Bk-Db-M-R
13/07	Silver Anniversary	15,283	Alloy	Bk-O-R
19	Black	4,573	Silver	Bk-Lb-M-O-R
19/47	Black/Silver	6,502	Alloy	Sv
26	Corvette Light Blue	1,960	Silver	Db
52	Corvette Yellow	1,243	Silver	Bk-Dbr-O
59	Corvette Light Beige	1,686	Silver	Bk-Db-Dbr-Lb-M
72	Corvette Red	2,074	Silver	Bk-Lb-O-R
82	Corvette Mahogany	2,121	Silver	Bk-Dbr-Lb-M-O
83	Corvette Dark Blue	2,084	Silver	Db-Lb-O
89	Corvette Dark Brown	1,991	Silver	Dbr-Lb-O

• Suggested interiors shown. Additional combinations were possible.

• Paint quantities exceed units sold. Sixteen units had primer only.

• Interior color quantities were 2,226 oyster cloth, 8,999 oyster leather, 5,703 silver leather, 1,353 black cloth, 7,218 black leather, 1,168 dark blue cloth, 3,207 dark blue leather, 693 light beige cloth, 3,338 light beige leather, 564 dark brown cloth, 2,449 dark brown leather, 1,221 red cloth, 4,607 red leather, 763 mahogany cloth, 2,443 mahogany leather.

Interior Codes: 12C=O/C, 122=O/L, 15C=Sv/C, 152=Sv/L, 19C=Bk/C, 192=Bk/L, 29C=Db/C, 292=Db/L, 59C=Lb/C, 592=Lb/L, 69C=Dbr/C, 692=Dbr/L, 72C=R/C, 722=R/L, 76C=M/C, 762=M/L.

Abbreviations: Bk=Black, C=Cloth, Db=Dark Blue, Dbr=Dark Brown, L=Leather, Lb=Light Beige, M=Mahogany, O=Oyster, R=Red, Sv=Silver.

1979 Corvette

Production: 53,807 coupes

1979 Numbers

Vehicle: 1Z8789S400001 through 1Z8789S453807
• Fifth digit varies as follows: 8=350ci, 195hp
4=350ci, 225hp

Suffix: ZAA: 350ci, 195hp, mt, ep   ZAH: 350ci, 195hp, at
ZAB: 350ci, 195hp, at, ep   ZAJ: 350ci, 195hp, at, ce
ZAC: 350ci, 195hp, at, ce, ep   ZBA: 350ci, 225hp, mt
ZAD: 350ci, 195hp, at, ha   ZBB: 350ci, 225hp, at
ZAF: 350ci, 195hp, mt

Block: 3970010: All   14016379: All (late production)

Head: 462624: All

Carb: Rochester Q-jet #17059202, 17059217: 350ci, 195hp, at
Rochester Q-jet #17059203: 350ci, 195hp, mt
Rochester Q-jet #17059210: 350ci, 225hp, at
Rochester Q-jet #17059211: 350ci, 225hp, mt
Rochester Q-jet #17059216: 350ci, 195hp, at, ac
Rochester Q-jet #17059228: 350ci, 225hp, at, ac
Rochester Q-jet #17059502: 350ci, 195hp, at, ce
Rochester Q-jet #17059504, 17059507: 350ci, 195hp, at, ac, ce
Rochester Q-jet #17059582: 350ci, 195hp, at, ha
Rochester Q-jet #17059584: 350ci, 195hp, at, ac, ha

Distributor: 1103285: 350ci, 195hp, ce   1103302: 350ci, 195hp
1103291: 350ci, 225hp   1103353: 350ci, 195hp

Alternator: 1101041, 1102394, 1102484: 350ci
1102474, 1102908: 350ci, ac

Abbreviations: ac=air conditioning, at=automatic transmission, ce=california emissions, ci=cubic inch, ep=early production, ha=high altitude, hp=horsepower, mt=manual transmission.

1979 Facts

• The new "high back" seat style introduced in the 1978 pace car replicas was made standard equipment in 1979. Extensive use of plastic in the frame structure resulted in weight reduction of about twelve pounds per seat. The new seats had better side bolster support, and the backs folded at a higher point to permit easier rear storage access. Inertia locking mechanisms restrained the backs in sudden deceleration, negating the need for manual locks. Driver and passenger seat tracks were redesigned for an additional inch of forward travel.

• The 1979 fuel filler pipe was redesigned to make it more difficult to modify for leaded-fuel access.

• Output of both the base L48 and optional L82 engines increased by 5hp due to a new "open flow" muffler design. Also, adding the L82's low restriction, dual-snorkel air intake to the base engine added another 5hp. The base L48 was rated at 195hp, the optional L82 at 225hp.

• The 85-mph speedometers required for 1980 models were used in some late-build 1979 production.

• In 1979, the previously optional AM-FM radio became standard equipment, and an illuminated visor-mirror combination became optional for the passenger side.

• Unlike 1978, 1979s equipped with manual transmissions had the same shock absorber rates as those with automatic transmissions.

• The front and rear spoilers developed as part of the 1978 pace car package became 1979 options. They were functional, decreasing drag by about 15% and increasing fuel economy by about a half-mile per gallon. Also, mirror-tint roof panels (RPO CC1), included in the pace car package and optional for late 1978s, were available as 1979 options.

• Tungsten-halogen headlight beams were phased into 1979 production early in the production year for increased visibility. These replaced only the high-beam units.

1979 Options

RPO #	DESCRIPTION	QTY	RETAIL $
1YZ87	Base Corvette Sport Coupe	53,807	$10,220.23
A31	Power Windows	20,631	141.00
AU3	Power Door Locks	9,054	131.00
CC1	Removable Glass Roof Panels	14,480	365.00
C49	Rear Window Defogger	41,587	102.00
C60	Air Conditioning	47,136	635.00
D35	Sport Mirrors	48,211	45.00
D80	Spoilers, front and rear	6,853	265.00
FE7	Gymkhana Suspension	12,321	49.00
F51	Heavy Duty Shock Absorbers	2,164	33.00
G95	Optional Rear Axle Ratio	428	19.00
K30	Cruise Control	34,445	113.00
L82	350ci, 225hp Engine	14,516	565.00
M21	4-Speed Manual Trans, close ratio	4,062	0.00
MX1	Automatic Transmission	41,454	0.00
NA6	High Altitude Emission Equipment	56	35.00
N37	Tilt-Telescopic Steering Column	47,463	190.00
N90	Aluminum Wheels (4)	33,741	380.00
QBS	White Letter SBR Tires, P255/60R15	17,920	226.20
QGR	White Letter SBR Tires, P225/70R15	29,603	54.00
U58	AM-FM Radio, stereo	9,256	90.00
UM2	AM-FM Radio, stereo with 8-track	21,435	228.00
UN3	AM-FM Radio, stereo with cassette	12,110	234.00
UP6	AM-FM Radio, stereo with CB	4,483	439.00
U75	Power Antenna	35,730	52.00
U81	Dual Rear Speakers	37,754	52.00
UA1	Heavy Duty Battery	3,405	21.00
YF5	California Emission Certification	3,798	83.00
ZN1	Trailer Package	1,001	98.00
ZQ2	Power Windows and Door Locks	28,465	272.00
ZX2	Convenience Group	41,530	94.00

• A 350ci, 195hp engine, 4-speed manual transmission, T-tops, and leather or cloth/leather interior trim were included in the base price.

• Base price and some options increased several times during 1979.

• RPO ZX2 included dome and courtesy light delay, headlight warning buzzer, underhood light, low fuel warning light, floor mats, intermittent wipers, and right side visor mirror.

• Manual transmission and/or L82 not available with NA6 or YF5.

• M21 close ratio transmission was no cost, but required L82 engine.

• Sales to Canada for 1979 were 5,227.

• Total production of 53,807 is accurate, but all option and interior color quantities are understated by about 7% because they were counted through August 31, 1979, but production continued into September.

1979 Colors

CODE	EXTERIOR	QTY	WHEELS	INTERIORS
10	Classic White	8,629	Silver	Bk-Db-Dg-Lb-O-R
13	Silver	7,331	Silver	Bk-Db-Dg-O-R
19	Black	10,465	Silver	Bk-Lb-O-R
28	Corvette Light Blue	3,203	Silver	Bk-Db-O
52	Corvette Yellow	2,357	Silver	Bk-Lb-O
58	Corvette Dark Green	2,426	Silver	Bk-Dg-Lb-O
59	Corvette Light Beige	2,951	Silver	Bk-Db-Dg-Lb-R
72	Corvette Red	6,707	Silver	Bk-Lb-O-R
82	Corvette Dark Brown	4,053	Silver	Bk-Lb-O
83	Corvette Dark Blue	5,670	Silver	Bk-Db-Lb-O-R

• Suggested interiors shown. Additional combinations were possible.

• Code 82 Corvette Dark Brown may also be coded 67.

• Fifteen 1979 Corvettes had primer only.

• Interior color quantities for 1979 were 2,606 oyster cloth, 9,316 oyster leather, 9,420 black leather, 2,142 dark blue cloth, 4,768 dark blue leather, 517 dark green cloth, 907 dark green leather, 2,224 light beige cloth, 8,818 light beige leather, 9,181 red leather.

Interior Codes: 12C=O/C, 122=O/L, 192=Bk/L, 29C=Db/C, 292=Db/L, 49C=Dg/C, 492=Dg/L, 59C=Lb/C, 592=Lb/L, 722=R/L.

Abbreviations: Bk=Black, C=Cloth, Db=Dark Blue, Dg=Dark Green, L=Leather, Lb=Light Beige, O=Oyster, R=Red.

1980 Corvette
Production: 40,614 coupes

1980 Numbers

Vehicle: 1Z878AS400001 through 1Z878AS440614
- Fifth digit varies as follows: 8=350ci,190hp H=305ci,180hp
                                   6=350ci, 230hp

Suffix: ZAK: 350ci, 190hp, at   ZBD: 350ci, 230hp, uu
        ZAM: 350ci, 190hp, mt   ZCA: 305ci, 180hp, at, ce
        ZBC: 350ci, 230hp, at

Block: 14010207: 350ci, 190hp, 230hp, lp   4715111: 305ci, 180hp
        3970010: 350ci, 190hp, 230hp

Head: 462624: 350ci, 190hp, 230hp   14014416: 305ci, 180hp

Carb: Rochester Q-jet #17080204: 350ci, 190hp, at
       Rochester Q-jet #17080207: 350ci, 190hp, mt
       Rochester Q-jet #17080228: 350ci, 230hp, at
       Rochester Q-jet #17080504, 17080517: 305ci, 180hp, at, ce

Distributor: 1103287: 350ci, 190hp, mt   1103368: 305ci, 180hp, at, ce
              1103352: 350ci, 190hp, at   1103435: 350ci, 230hp, at
              1103353: 350ci, 190hp, at

Alternator: 1101041, 1101075, 1101085, 1101088, 1103122

Ending Vehicle:
Sep 79: 400011	Jan 80: 416198	May 80: 431152
Oct 79: 404267	Feb 80: 420057	Jun 80: 434509
Nov 79: 408343	Mar 80: 424380	Jul 80: 438049
Dec 79: 411652	Apr 80: 427800	Aug 80: 440614

Abbreviations: at=automatic transmission, ce=california emissions, ci=cubic inch, hp=horsepower, lp=late production, mt=manual transmission, uu=uncertain usage.

1980 Facts

• For the first time since 1974, two engine displacements were available in Corvettes, but not optional. Because of tightened California emission restrictions, Chevrolet did not certify its 350 cubic-inch engines there in 1980. California Corvette buyers were required to purchase a 305ci engine, the RPO LG4, with a $50 credit. This was a standard passenger car engine. In Corvettes, it could be combined only with automatic transmissions.

• The 4-speed manual transmission was not available with the optional L82 engine, with the possible exception of a few early production builds.

• This model featured new front and rear bumper "caps" with integral spoilers. Radiator air flow increased about fifty-percent. Integrated spoilers improved the drag coefficient from .503 to .443 compared to the 1979 model equipped with optional, non-integrated spoilers.

• The crossed flag emblems for 1980 were a new, more elongated design.

• The speedometers for 1980 Corvettes read to a maximum of 85 mph, a new federal requirement. These were phased in during 1979 production.

• The rear storage compartments were changed from a three-lid design to a two-lid style in the 1980 model. The battery stayed in a separate compartment behind the driver, but the center and passenger-side compartments were combined with one access door.

• After several years of weight increases, the Corvette was lighter in 1980 as engineers trimmed weight by using lower density roof panels, reducing the thickness of hood and outer doors, and using aluminum for the differential housing and crossmember. For the base L48 engine, the L82's aluminum intake manifold became standard.

• For improved surface quality, urethane was applied during molding to panels forward of the firewall and to roof panels.

• The California RPO LG4 305ci engine included stainless-steel tubular exhaust headers with an oxygen sensor in a "closed loop" system. Despite its lower displacement and more restrictive emissions equipment, the Corvette application of this engine developed 180hp, just 10hp less than the 350ci, L48 base engine for other states.

1980 Options

RPO#	DESCRIPTION	QTY	RETAIL $
1YZ87	Base Corvette Sport Coupe	40,614	$13,140.24
AU3	Power Door Locks	32,692	140.00
CC1	Removable Glass Roof Panels	19,695	391.00
C49	Rear Window Defogger	36,589	109.00
FE7	Gymkhana Suspension	9,907	55.00
F51	Heavy Duty Shock Absorbers	1,695	35.00
K30	Cruise Control	30,821	123.00
LG4	305ci, 180hp Engine (required in California)	3,221	-50.00
L82	350ci, 230hp Engine	5,069	595.00
MM4	4-Speed Manual Transmission	5,726	0.00
MX1	Automatic Transmission	34,838	0.00
N90	Aluminum Wheels (4)	34,128	407.00
QGB	White Letter SBR Tires, P225/70R15	26,208	62.00
QXH	White Letter SBR Tires, P255/60R15	13,140	426.16
UA1	Heavy Duty Battery	1,337	22.00
U58	AM-FM Radio, stereo	6,138	46.00
UM2	AM-FM Radio, stereo with 8-track	15,708	155.00
UN3	AM-FM Radio, stereo with cassette	15,148	168.00
UP6	AM-FM Radio, stereo with CB	2,434	391.00
U75	Power Antenna	32,863	56.00
UL5	Radio Delete	201	-126.00
U81	Dual Rear Speakers	36,650	52.00
V54	Roof Panel Carrier	3,755	125.00
YF5	California Emission Certification	3,221	250.00
ZN1	Trailer Package	796	105.00

• A 350ci, 190hp, (305ci, 180hp in California for $50 credit), 4-speed manual transmission or automatic transmission, T-tops, and leather/vinyl or cloth/vinyl interior trim were included in the base price.

• The Corvette's base price increased four times during 1980, increasing from $13,140.24 to $14,345.24. Option prices were not affected.

• RPO A31 power windows, RPO C60 air conditioning, and RPO N37 tilt-telescopic steering column, all optional for part of 1979, were included in the 1980's base price.

• RPO C49 rear window defogger included UA1 heavy-duty battery.

• RPO LG4 305ci, 180hp engine was required for California and not available elsewhere. It was not available with manual transmission.

• RPO V54 roof panel carrier mounted to the rear deck for external transport of removable roof panels. It was new for 1980.

• RPO ZN1 trailer package included heavy-duty radiator, not available as a separate option, and RPO FE7 gymkhana suspension.

• Sales to Canada for 1980 were 5,020.

1980 Colors

CODE	EXTERIOR	QTY	WHEELS	INTERIORS
10	White	7,780	Silver	Bk-Cl-Db-Ds-O-R
13	Silver	4,341	Silver	Bk-Cl-Db-O-R
19	Black	7,250	Silver	Bk-Ds-O-R
28	Dark Blue	4,135	Silver	Bk-Db-Ds-O-R
47	Dark Brown	2,300	Silver	Bk-Ds-O
52	Yellow	2,077	Silver	Bk-O
58	Dark Green	844	Silver	Bk-Ds-O
59	Frost Beige	3,070	Silver	Bk-Cl-Db-Ds-R
76	Dark Claret	3,451	Silver	Bk-Cl-Ds-O
83	Red	5,714	Silver	Bk-Ds-O-R

• Suggested interiors shown. Additional combinations were possible.

• Paint quantites exceed actual units sold. Records show no primer only, but additional special colors coded "spec" were used.

• Code 492 for green leather interior was released, but cancelled early.

• Interior color quantities for 1980 were 1,808 oyster cloth, 7,104 oyster leather, 6,776 black leather, 1,040 dark blue cloth, 2,752 dark blue leather, 1,616 doeskin cloth, 7,310 doeskin leather, 7,184 red leather, 1,338 claret cloth, 3,636 claret leather.

Interior Codes: 12C=O/C, 122=O/L, 192=Bk/L, 29C=Db/C, 292=Db/L, 59C=Ds/C, 592=Ds/L, 722=R/L, 79C=CL/C, 792=CL/L.

Abbreviations: Bk=Black, C=Cloth, CL=Claret, Db=Dark Blue, Ds=Doeskin, L=Leather, O=Oyster, R=Red.

1981 Corvette

Production: 40,606 coupes

1981 Numbers

Vehicle: 1G1AY8764BS400001 thru 1G1AY8764BS431611 (St Louis)
1G1AY8764B5100001 thru 1G1AY8764B5108995 (B-Green)
• Ninth digit is a security code and varies.

Suffix: ZDA: 350ci, 190hp, mt   ZDC: 350ci, 190hp, mt, ce
ZDB: 350ci, 190hp, at, ce   ZDD: 350ci, 190hp, at

Block: 14010207: All

Head: 462624: All

Carburetor: Rochester Q-jet #17081217: 350ci, 190hp, mt
Rochester Q-jet #17081218: 350ci, 190hp, at, ce
Rochester Q-jet #17081228: 350ci, 190hp, at

Distributor: 1103443: All

Alternator: 1101075, 1101085, 1103088, 1103091, 1103103

Ending Vehicle (St. Louis)		
Aug 80: 400775	Dec 80: 415234	Apr 81: 426422
Sep 80: 404136	Jan 81: 418399	May 81: 428003
Oct 80: 408594	Feb 81: 421392	Jun 81: 429775
Nov 80: 412124	Mar 81: 424742	Jul 81: 431611

Abbreviations: at=automatic transmission, ce=california emissions, ci=cubic inch, hp=horsepower, mt=manual transmission.

1981 Facts

• The 1981 Corvette was the first model year to be built simultaneously in two locations. The first Corvette was completed at the new Bowling Green, Kentucky, assembly line on June 1, 1981. The last Corvette built at St. Louis was completed on August 1, 1981.

• Although there were no engine options for 1981, the base 350ci, 190hp L81 engine was certified for sale in all states including California with both 4-speed manual and automatic transmissions.

• Exterior styling carried over from 1980, but emblems did change slightly.

• The tubular stainless steel exhaust manifolds, used for 1980 Corvettes sold in California with the LG4 305ci engine, were standard with the 1981 base engine.

• Chevrolet's "computer command control" used on 1980 Corvettes sold in California became standard equipment on all 1981 Corvettes. The system automatically adjusted ignition timing and air-fuel mixture.

• Chevrolet introduced a fiberglass-reinforced monoleaf rear spring for 1981 Corvettes equipped with automatic transmissions and standard suspensions. The plastic spring weighed eight pounds compared to forty-four pounds for the steel unit it replaced.

• The anti-theft alarm system was improved in 1981 by the addition of an ignition interrupt to prevent engine start.

• All 1981 valve covers were magnesium for weight reduction.

• For improved fuel economy, 1981 Corvettes with automatic transmissions had torque converter clutches for second and third gears.

• A detail change to the 1981 Corvette interior was the color-keying of the headlamp and windshield wiper switch bezels to the interior color. In 1980, they were black regardless of interior color.

• The 1981 Corvette was the last model to have a manual transmission available until well into the 1984 production year.

• The St. Louis assembly plant used mainly lacquer paints through the end of production in that facility, while Bowling Green used enamel basecoats followed by clear topcoats. Bowling Green painted two-tones and solids. St. Louis painted mainly solids, but a limited two-tone production run at St. Louis of about 100 cars using the Bowling Green enamel-clearcoat system has been verified. These were first offered to GM employees and may not have been available to retail customers.

• A power seat became available in Corvettes for the first time in 1981 as RPO A42. It was available for the driver's side only.

1981 Options

RPO#	DESCRIPTION	QTY	RETAIL $
1YY87	Base Corvette Sport Coupe	40,606	$16,258.52
AU3	Power Door Locks	36,322	145.00
A42	Power Driver Seat	29,200	183.00
CC1	Removable Glass Roof Panels	29,095	414.00
C49	Rear Window Defogger	36,893	119.00
DG7	Electric Sport Mirrors	13,567	117.00
D84	Two-Tone Paint	5,352	399.00
FE7	Gymkhana Suspension	7,803	57.00
F51	Heavy Duty Shock Absorbers	1,128	37.00
G92	Performance Axle Ratio	2,400	20.00
K35	Cruise Control	32,522	155.00
MM4	4-Speed Manual Transmission	5,757	0.00
N90	Aluminum Wheels (4)	36,485	428.00
QGR	White Letter SBR Tires, P225/70R15	21,939	72.00
QXH	White Letter SBR Tires, P255/60R15	18,004	491.92
UL5	Radio Delete	315	-118.00
UM4	AM-FM Radio, etr stereo with 8-track	8,262	386.00
UM5	AM-FM Radio, etr stereo with 8-track/CB	792	712.00
UM6	AM-FM Radio, etr stereo with cassette	22,892	423.00
UN5	AM-FM Radio, etr stereo with cassette/CB	2,349	750.00
U58	AM-FM Radio, stereo	5,145	95.00
U75	Power Antenna	32,903	55.00
V54	Roof Panel Carrier	3,303	135.00
YF5	California Emission Certification	4,951	46.00
ZN1	Trailer Package	916	110.00

• A 350ci, 190hp engine, 4-speed manual transmission or automatic transmission, T-tops, and leather/vinyl, or cloth/vinyl interior trim were included in the base price.

• All optional radios except U58 were new-style Delcos with electronic tuned receivers (etr). Available with 8-track, 8-track plus CB, cassette, or cassette/CB, the radios featured digital station tuning readout and built-in clocks. If a 1981 Corvette had one of these radios, the standard quartz instrument panel clock was replaced with an oil temperature gauge.

1981 Colors

CODE	EXTERIOR	QTY	PLANT	INTERIORS
06	Mahogany Metallic	1,092	St. Louis	Cm-Dr
10	White	6,387	St. Louis	Bk-Ch-Cm-Db-Dr-Mr-Sg
13	Silver Metallic	2,590	St. Louis	Bk-Ch-Db-Mr-Sg
19	Black	4,712	St. Louis	Bk-Ch-Cm-Db-Dr-Mr-Sg
24	Bright Blue Metallic	1	St. Louis	Ch-Cm-Db-Sg
28	Dark Blue Metallic	2,522	St. Louis	Cm-Db-Mr-Sg
33	Silver Metallic	3,369	B. Green	Ch-Db-Dr-Sg
38	Dark Blue Metallic	496	B. Green	Cm-Db-Sg
39	Charcoal Metallic	613	B. Green	Ch-Dr-Sg
50	Beige	2,239	B. Green	Ch-Cm-Db-Dr
52	Yellow	1,031	St. Louis	Bk-Ch-Cm
59	Beige	3,842	St. Louis	Cm-Db-Dr-Mr
74	Dark Bronze	432	B. Green	Ch-Cm
75	Red	4,310	St. Louis	Bk-Ch-Cm-Mr-Sg
79	Maroon Metallic	1,618	St. Louis	Bk-Ch-Cm-Mr-Sg
80	Autumn Red	1,505	B. Green	Ch-Cm-Dr-Sg
84	Charcoal Metallic	3,485	St. Louis	Bk-Ch-Cm-Mr-Sg
98	Dark Claret Metallic	341	B. Green	Cm-Dr-Sg
33/38	Silver/Dark Blue	–	St.L/BG	Db-Sg
33/39	Silver/Charcoal	–	St.L/BG	Ch-Sg
50/74	Beige/Dark Bronze	–	St.L/BG	Cm
80/98	Autumn Red/Dark Claret	–	St.L/BG	Dr-Sg

• Suggested interiors shown. Other combinations were possible.

• Two-tone quantities were included in the single code counts. One primer 1981 is known, a St. Louis build, with "SPEC" trim tag.

• Two-tones (about 25 of each combination) were also built at St. Louis in late 1980, prior to production startup at Bowling Green. At least some trim plates for these had only the lower body color code.

• **Interior Codes:** 13C=Sg/C, 132=Sg/L, 152=Sg/L, 182=Ch/L, 19C=Bk/C, 192=Bk/L, 29C=Db/C, 292=Db/L, 64C=Cm/C, 642=Cm/L, 67C=Dr/C, 672=Dr/L, 74C=Dr/C, 742=Dr/L, 752=Mr/L.

• **Abbreviations:** Bk=Black, C=Cloth, Ch=Charcoal, Cm=Camel, Db=Dark Blue, Dr=Dark Red, L=Leather, Mr=Medium Red, Sg=Silver Gray.

1982 Corvette

1982 Numbers

Vehicle: 1G1AY8786C5100001 thru 1G1AY8786C5125408
- Sixth digit is a zero for the Collector Edition
- Ninth digit is a security code and varies.
- Vin tag ending 00017 was lost and not built.

Suffix: ZBA: 350ci, 200hp, at   ZBN: 350ci, 200hp, at, ce
ZBC: 350ci, 200hp, at, ce, ep

Block: 14010207: All

Head: 462624: All

Throttle Body Injection: 17082052: Rear Unit   17082053: Front Unit

Distributor: 1103479: All

Alternator: 1101071, 1101075, 1103091, 1103103

Ending Vehicle:

Oct 81: 100564	Mar 82: 110060	Jul 82: 120227
Nov 81: 100590	Apr 82: 112598	Aug 82: 121889
Dec 81: 102647	May 82: 115020	Sep 82: 124433
Jan 82: 105004	Jun 82: 117686	Oct 82: 125408
Feb 82: 107287		

Abbreviations: at=automatic transmission, ce=california emissions, ci=cubic inch, ep=early production, hp=horsepower.

1982 Facts

• The 1982 Corvette was the last of a generation which dated its basic body structure to 1968 and its chassis to 1963. To honor the 1982 model's special status, Chevrolet offered a "Collector Edition." In addition to a higher level of standard features optional on base models, the Collector Edition had a lifting rear hatchback glass, a Corvette first, special wheels styled to resemble 1967's "bolt on" optional wheels, unique silver-beige paint, silver-beige leather interior, and special cloisonne emblems.

• The Collector Edition Hatchback Coupe carried a special code (zero in the sixth digit) in its vehicle identification number, but didn't have a separate serial sequence. At $22,537.59, it was the first Corvette with a base price exceeding $20,000.00.

• A manual transmission was not available in 1982 Corvettes.

• The automatic transmission in 1982 Corvettes was a new four-speed unit with a torque converter clutch operating in the top three gears. It used a higher first gear ratio (3.07:1) for improved acceleration.

• Hoods of 1982 models had solenoid-operated doors to direct fresh air directly into the air cleaner during full throttle.

• Chevrolet introduced "cross fire injection" on the 1982 Corvette. This wasn't fuel injection of the type available in 1957-1965 Corvettes; rather, it combined two "injectors" with Chevrolet's Computer Command Control system to achieve better economy, driveability and performance through more precise metering of the fuel. The Computer Command Control itself was refined in 1982 so that it was capable of making eighty adjustments per second compared to ten the previous year.

• The new fuel metering system used in 1982 included a positive fuel cutoff to prevent engine run-on (dieseling).

• The charcoal air filtering element of the 1981 model was replaced with a paper element in the cross-fire-injection 1982.

• The exhaust system of 1982 models was redesigned with a smaller and lighter catalytic converter. The exhaust pipes leading into the converter were redesigned to deliver hotter exhaust gases to the converter to increase its efficiency.

• All 1982 Corvettes were built in the new Corvette plant in Bowling Green, Kentucky. Production was initiated in 1981 when 8,995 models were built. Production for 1982 was 25,407.

• The 1982 Corvette was the last model with optional radio packages that included an 8-track tape (RPO UM4).

1982 Options

RPO #	DESCRIPTION	QTY	RETAIL $
1YY87	Base Corvette Sport Coupe 18,648		$18,290.07
1YY07	Corvette Collector Edition Hatchback 6,759		22,537.59
AG9	Power Driver Seat 22,585		197.00
AU3	Power Door Locks 23,936		155.00
CC1	Removable Glass Roof Panels 14,763		443.00
C49	Rear Window Defogger 16,886		129.00
DG7	Electric Sport Mirrors 20,301		125.00
D84	Two-Tone Paint 4,871		428.00
FE7	Gymkhana Suspension 5,457		61.00
K35	Cruise Control 24,313		165.00
N90	Aluminum Wheels 16,844		458.00
QGR	White Letter SBR Tires, P225/70R15 5,932		80.00
QXH	White Letter SBR Tires, P255/60R15 19,070		542.52
UL5	Radio Delete 150		-124.00
UM4	AM-FM Radio, etr stereo with 8-track 923		386.00
UM6	AM-FM Radio, etr stereo with cassette 20,355		423.00
UN5	AM-FM Radio, etr stereo with cassette/CB ... 1,987		755.00
U58	AM/FM Radio, stereo 1,533		101.00
U75	Power Antenna 15,557		60.00
V08	Heavy Duty Cooling 6,006		57.00
V54	Roof Panel Carrier 1,992		144.00
YF5	California Emission Certification 4,951		46.00

• A 350ci, 200hp engine, automatic transmission, T-tops, and leather/vinyl or cloth/vinyl interior trim were included in the base price.

• There were no optional Corvette engines in 1982.

• Manual transmissions were not available.

• Corvette Collector Edition Hatchback Coupe included RPOs CC1, C49, QXH, U75, special silver-beige paint, graduated hood and side body decals, commemorative aluminum wheels, frameless glass hatchback with manual remote release, accent pinstriping, multi-tone silver-beige leather seats and door trim, leather wrapped steering wheel and horn cap, cloisonne exterior and interior emblems, and luxury carpeting. If UN5 radio was selected, it cost $695 instead of $755.

• RPO N90 wheels were different than Collector Edition alloys.

• Sales to Canada for 1982 were 459. Sales to Japan were 9.

1982 Colors

CODE	EXTERIOR	QTY	WHEELS	INTERIORS
10	White	2,975	Silver	Ch-Cm-Db-Dr-Sgn-Sgy
13	Silver	711	Silver	Ch-Db-Dr-Sgy
19	Black	2,357	Silver	Ch-Cm-Db-Sgn-Sgy
24	Silver Blue	1,124	Silver	Ch-Cm-Sgy
26	Dark Blue	562	Silver	Cm-Db-Sgy
31	Bright Blue....................	567	Silver	Ch-Cm-Db-Sgy
39	Charcoal	1,093	Silver	Ch-Dr-Sgy
40	Silver Green	723	Silver	Ch-Sgn
56	Gold	648	Silver	Ch-Cm
59	Silver Beige	6,759	Alloy	Sb
70	Red	2,155	Silver	Ch-Cm-Dr-Sgy
99	Dark Claret	853	Silver	Cm-Dr-Sgy
10/13	White/Silver	664	Silver	Ch-Sgy
13/39	Silver/Charcoal	1,239	Silver	Ch-Dr-Sgy
13/99	Silver/Dark Claret	1,301	Silver	Dr-Sgy
24/26	Silver Blue/Dark Blue	1,667	Silver	Db-Sgy

• Suggested interiors shown. Other combinations were possible.

• Nine 1982 Corvettes had primer only.

• Exterior code 59 Silver-Beige, and interior code 592 silver-beige leather were exclusive to the Collector Edition Hatchback.

• Interior color quantities for 1982 were 992 silver gray cloth, 3,083 silver gray leather, 2,706 charcoal leather, 451 dark blue cloth, 2,750 dark blue leather, 757 silver green leather, 6,759 silver beige leather (collector), 350 camel cloth, 1,710 camel leather, 905 red cloth, 4,444 red leather.

Interior Codes: 13C=Sgy/C, 132=Sgy/L, 182=Ch/L, 22C=Db/C, 222=Db/L, 402=Sgn/L, 592=Sb/L, 64C=Cm/C, 642=Cm/L, 74C=Dr/C, 742=Dr/L.

Abbreviations: C=Cloth, Ch=Charcoal, Cm=Camel, Db=Dark Blue, Dr=Dark Red, L=Leather, Sb=Silver Beige, Sgn=Silver Green, Sgy=Silver Gray.

1984 Corvette

Production: 51,547 coupes

1984 Numbers

Vehicle: 1G1AY0782E5100001 thru 1G1AY0782E5151547
  • Ninth digit is a security code and varies.

Suffix:

ZFC: 350ci, 205hp, at	ZFK: 350ci, 205hp, at
ZFD: 350ci, 205hp, mt	ZFL: 350ci, 205hp, mt
ZFF: 350ci, 205hp, at, ce	ZFM: 350ci, 205hp, at, ce
ZFH: 350ci, 205hp, ex	ZFN: 350ci, 205hp, mt, ce
ZFJ: 350ci, 205hp, mt, ex	ZFR: 350ci, 205hp, mt, ce

Block: 14010207: All

Head: 462624: All

Abbreviations: at=automatic transmission, ce=california emissions, ci=cubic inch, ex=export, hp=horsepower, mt=manual transmission.

1984 Facts

• The 1984 Corvette was a complete redesign in almost every aspect. Handling considerations dominated and the result was praised by the motoring press as the world's best cornering automobile.

• The 1984 Corvette was introduced in March 1983. Because of its late release and since it met all 1984 federal requirements, Chevrolet general manager Robert Stemple decided to bypass the 1983 model designation. The result was a very long production run and the second highest model year volume in the Corvette's history. 1983 Corvettes were built, serial numbered, and tested by both Chevrolet and the motoring press at the "long lead" preview at Riverside Raceway in December 1982. But 1983 Corvettes were not released for public sale.

• Design criteria specified that the 1984 Corvette have more ground clearance but less overall height, a lower center of gravity, and better front-to-rear weight distribution. In order to achieve these goals, engineers located the engine more rearward, then routed the exhaust system through a much larger transmission tunnel. The effects on handling were dramatic, but a penalty was paid in interior room, especially in the footwell areas.

• A "4+3", 4-speed manual transmission, built by the Doug Nash Company, had overdrives in the top three gears for improved fuel economy. It was not available early.

• All 1984 Corvettes were designed with one-piece, lift-off roof panels and rear hatch windows. At the time, the rear window glass was the largest compound glass ever installed in an American automobile. The front windshield was raked at the greatest angle, 64°.

• Brakes remained disc at all four wheels, but components were new and included aluminum calipers supplied by Girlock of Australia.

• Electronic instrumentation was standard and included digital readouts for engine monitoring and liquid crystal graphic displays for speed and engine revolutions. Analog instrumentation was not available.

• The 1984 Corvette was designed with a pad protruding from the passenger side of the dash. This was part of a passive restraint system conceived when it was assumed that federal regulation would require such restraints. The Reagan Administration dropped the restraint proposals, but the Corvette's pad remained.

• The 1984 Corvette was designed without fiberglass seams on exposed panels to eliminate factory finishing. The exterior seams were under the rub molding extending around the entire body.

• The radiator was a new design using aluminum for the cooling fins and plastic for the reservoirs. A thermostatically controlled electric fan operated only when needed and only under 35mph.

• Chevrolet built specially modified 1984 Corvettes for the export markets of European, Middle East, Japanese, and Latin American countries. Changes included different license plate provisions, leaded fuel capability, and electrical, glass, lighting and mirror modifications.

• Single transverse plastic leaf springs were used front and rear.

1984 Options

RPO #	DESCRIPTION	QTY	RETAIL $
1YY07	Base Corvette Sport Coupe	51,547	$21,800.00
AG9	Power Driver Seat	48,702	210.00
AQ9	Sport Seats, cloth	4,003	625.00
AR9	Base Seats, leather	40,568	400.00
AU3	Power Door Locks	49,545	165.00
CC3	Removable Transparent Roof Panel	15,767	595.00
D84	Two-Tone Paint	8,755	428.00
FG3	Delco-Bilstein Shock Absorbers	3,729	189.00
G92	Performance Axle Ratio	410	22.00
KC4	Engine Oil Cooler	4,295	158.00
K34	Cruise Control	49,832	185.00
MM4	4-Speed Manual Transmission	6,443	0.00
QZD	P255/50VR16 Tires/16" Wheels	51,547	561.20
UL5	Radio Delete	104	-331.00
UM6	AM-FM Stereo Cassette	6,689	153.00
UN8	AM-FM Stereo, Citizens Band	178	215.00
UU8	Stereo System, Delco-Bose	43,607	895.00
V01	Heavy-Duty Radiator	12,008	57.00
YF5	California Emission Requirements	6,833	75.00
Z51	Performance Handling Package	25,995	600.20
Z6A	Rear Window+Side Mirror Defoggers	47,680	160.00

• A 350ci, 205hp engine, 4-speed automatic transmission, removable body-color roof panel, and cloth seats were included in the base price.

• Optional leather seats were the same design as the base cloth style. Sport seats were available in cloth (different material than base) and featured inflatable lumbar support and power-adjusted side bolsters.

• The RPO QZD 16-inch tire and wheel package, initially intended to be included as part of the Z51 option, and as a separate option for base models, was required for all 1984 Corvettes sold. The 15-inch alloy wheels and P215/65R15 tires appeared in dealer order guides but were not used.

• RPO Z51 included heavy-duty front and rear springs, shock absorbers, stabilizer bars and bushings, fast steering ratio, engine oil cooler, extra radiator fan (pusher), P255/50VR16 tires and directional alloy wheels, (16x8.5-inch front, 16x9.5-inch rear).

• The vehicle base price increased after the start of production to reflect standard 16-inch wheels, and the cost of Z51 was reduced.

• Sales to Canada for 1984 were 1,759.

1984 Colors

CODE	EXTERIOR	QTY	WHEELS	INTERIORS
10	White	6,416	Alloy	Ca-Br-Gr-Mb-Mg-S
16	Bright Silver Metallic	3,109	Alloy	Gr-Mg
18	Medium Gray Metallic	3,147	Alloy	Gr-Mg
19	Black	7,906	Alloy	Ca-Gr-Mg-S
20	Light Blue Metallic	1,196	Alloy	Mb
23	Medium Blue Metallic	1,822	Alloy	Mb
35	Yellow	1	Alloy	Gr
53	Gold Metallic	2,430	Alloy	S
63	Light Bronze Metallic	2,452	Alloy	Br
66	Dark Bronze Metallic	1,371	Alloy	Br
72	Bright Red	12,942	Alloy	Gr-S
16/18	Silver/Medium Gray	3,629	Alloy	Gr-Mg
20/23	Light Blue/Medium Blue	1,433	Alloy	Mb
63/66	Light Bronze/Dark Bronze	3,693	Alloy	Br

• Additional codes: 70 and 33 for Bright Red, 41 for Black, 40 for White.

• Suggested interiors shown. Other combinations were possible.

• Interior colors sold in 1984 were 1,580 graphite cloth, 1,251 graphite sport cloth, 10,921 graphite leather, 806 gray cloth, 594 gray sport cloth, 5,141 gray leather, 2,858 blue cloth, 924 blue sport cloth, 684 saddle cloth, 548 saddle sport cloth, 5,453 saddle leather, 1,048 bronze cloth, 686 bronze sport cloth, 6,285 bronze leather, 12,768 red leather.

• All wheels were alloy with similar exterior appearance. Base models had all 16x8.5-inch. Z51 models had 16x8.5-inch front, 16x9.5-inch rear.

Interior Codes: 12C=Gr/C, 12V=Gr/Sc, 122=Gr/L, 15C=Mg/C, 15V=Mg/Sc, 152=Mg/L, 28C=Mb/C, 28V=Mb/Sc, 62C=S/C, 62V=S/Sc, 622=S/L, 65C=Br/C, 65V=Br/Sc, 652=Br/L, 742=Ca/L.

Abbreviations: Ca=Carmine, Br=Bronze, C=Cloth, Gr=Graphite, L=Leather, Mb=Medium Blue, Mg=Medium Gray, S=Saddle, Sc=Sport Seat Cloth.

1985 Corvette

Production: 39,729 coupes

1985 Numbers

Vehicle: 1G1YY0787F5100001 thru 1G1YY0787F5139729
     • Ninth digit is a security code and varies.

Suffix: ZDF: 350ci, 230hp, at   ZJJ: 350ci, 230hp, at, oc
         ZJB: 350ci, 230hp, mt   ZJK: 350ci, 230hp, mt, oc
         ZJC: Export

Block: 14010207: All

Head: 462624: All

Abbreviations: at=automatic transmission, ci=cubic inch, hp=horsepower, mt=manual transmission, oc=engine oil cooler.

1985 Facts

• The 1982 and 1984 Corvette engines had "Cross Fire Injection," but genuine fuel injection returned to the Corvette in 1985 for the first time in two decades. The 1985 tuned-port injection, built by Bosch, was standard equipment and featured a mass airflow sensor, aluminum-tube tuned intake runners, a mold-cast plenum, and an air cleaner mounted forward of the radiator support. This new L98 engine delivered a horsepower increase from 205hp to 230hp, a torque increase from 290 lb.-ft. to 330, and a real-world fuel economy increase of about 11%.

• Suspension rates were lowered in 1985, a result of harsh ride criticism. Springs for the base suspension were softer by 26% in front, 25% in the rear. Springs for RPO Z51 were 16% softer in front and 25% softer in the rear. To compensate for the spring change, larger-diameter stabilizer bars were included with Z51-equipped models.

• The RPO Z51 suspension option included Delco-Bilstein shock absorbers and heavy-duty cooling, but these items were available separately with standard suspensions as RPOs FG3 and V08 respectively.

• In its January 1985 "ten best" issue, *Car and Driver* magazine pronounced the Corvette to be America's fastest production car at an even 150mph top speed. Corvette also took top honors in top-gear acceleration, and tied for best (with Porsche) in roadholding as measured by G-force skidpad adhesion of .84.

• The bore of the brake master cylinder was increased in 1985, and the booster itself was plastic, the first such application in an American car. The new plastic booster was 30% lighter and less subject to corrosion.

• Manual transmission 1985s came with a new, heavy-duty 8.5-inch ring differential. Rear axle gearing ratio for manual transmission models was 3.07:1. Standard gearing ratio for automatic transmissions was 2.73:1, but the 3.07:1 could be ordered as RPO G92.

• The overdrive selector button for manual transmissions was moved during 1985 production from a console location to the shift knob itself.

• Wheel balance weights changed in 1985 from the outside-rim, clip-on style, to an inner-surface adhesive type. The change was mainly for aesthetics, but Chevrolet also believed a better balance resulted because of the adhesive weight's proximity to the wheel's depth center.

• A full length oil pan gasket reinforcement was added to the 1985 Corvette engine to improve gasket compression seal.

• The 1985 Corvette distributor was modified to prevent distributor spark ignition of exterior fuel vapors.

• Electronic instrumentation continued much as the previous year, but displays were revised and improved with cleaner graphics, less color on the speedometer and tachometer, and larger digits for the center-cluster liquid crystal displays.

• The sport seat option which was available in cloth only in 1984, became available in leather as an interim 1985 option.

• A map strap was added to the 1985's driver-side sun shade.

• Electronic air conditioning was announced as a late 1985 option, but introduction was delayed into the 1986 model year.

1985 Options

RPO #	DESCRIPTION	QTY	RETAIL $
1YY07	Base Corvette Sport Coupe	39,729	$24,403.00
AG9	Power Driver Seat	37,856	215.00
AQ9	Sport Seats, leather	—	1,025.00
AR9	Base Seats, leather	—	400.00
—	Sport Seats, cloth	5,661	625.00
AU3	Power Door Locks	38,294	170.00
CC3	Removable Transparent Roof Panel	28,143	595.00
D84	Two-Tone Paint	6,033	428.00
FG3	Delco-Bilstein Shock Absorbers	9,333	189.00
G92	Performance Axle Ratio	5,447	22.00
K34	Cruise Control	38,369	185.00
MM4	4-Speed Manual Transmission	9,576	0.00
NN5	California Emission Requirements	6,583	99.00
UL5	Radio Delete	172	-256.00
UM6	AM-FM Stereo Cassette	2,958	122.00
UN8	AM-FM Stereo, Citizens Band	16	215.00
UU8	Stereo System, Delco-Bose	35,998	895.00
V08	Heavy-Duty Cooling	17,539	225.00
Z51	Performance Handling Package	14,802	470.00
Z6A	Rear Window+Side Mirror Defoggers	37,720	160.00

• A 350ci, 230hp engine, 4-speed automatic transmission, removable body-color roof panel, and cloth seats were included in the base price.

• After production started, sport seats became available in leather in addition to cloth Leather seats sold totaled 30,955, which included both the optional base seat in leather and the optional sport seat in leather.

• RPO Z51 included FG3 Delco-Bilstein shock absorbers, V08 heavy-duty cooling, heavy-duty front and rear springs, stabilizers and bushings, fast steering ratio and 16x9.5-inch wheels. FG3 and V08 were available separately with non-Z51 models.

• RPO CC3 roof panel was given stronger sun screening for 1985.

• Non-USA sales were 1,944 Canada, 33 Japan, 6 Europe, and 16 others.

1985 Colors

CODE	EXTERIOR	QTY	WHEELS	INTERIORS
13	Silver Metallic	1,752	Alloy	Gr-Mg
18	Medium Gray Metallic	2,519	Alloy	Gr-Mg
20	Light Blue Metallic	1,021	Alloy	Mb
23	Medium Blue Metallic	2,041	Alloy	Mb
40	White	4,455	Alloy	Ca-Gr-S-Mb-Mg
41	Black	7,603	Alloy	Ca-Gr-S-Mg
53	Gold Metallic	1,411	Alloy	S
63	Light Bronze Metallic	1,440	Alloy	Br
66	Dark Bronze Metallic	1,030	Alloy	Br
81	Bright Red	10,424	Alloy	Ca-Gr-S-Mg
13/18	Silver/Gray	2,170	Alloy	Gr-Mg
20/23	Light Blue/Medium Blue	1,470	Alloy	Mb
63/66	Light Bronze/Dark Bronze	2,393	Alloy	Br

• Suggested interiors shown. Other combinations were possible.

• Code 33 also used for Bright Red. Exterior colors for 1985 were the same as 1984, except Silver and Bright Red were brighter hues.

• Three 1985s were factory painted 1986 yellow. Trim labels in these cars were manually changed to reflect yellow's paint code 35.

• Interior colors sold in 1985 were 818 graphite cloth, 1,763 graphite sport cloth, 9,346 graphite leather, 571 gray cloth, 1,045 gray sport cloth, 3,579 gray leather, 957 blue cloth, 1,365 blue sport cloth, 2,121 blue leather, 349 saddle cloth, 607 saddle sport cloth, 3,759 saddle leather, 417 bronze cloth, 881 bronze sport cloth, 3,878 bronze leather, 8,272 red leather. Note that leather quantities include both base and sport seats.

• All wheels were alloy with similar exterior appearance. Standard wheels were 16x8.5-inch front/rear; Z51 had 16x9.5-inch wheels front/rear.

Interior Codes: 12C=Gr/C, 12V=Gr/Sc, 122=Gr/L, 15C=Mg/C, 15V=Mg/Sc, 152=Mg/L, 28C=Mb/C, 28V=Mb/Sc, 282=Mb/L, 62C=S/C, 62V=S/Sc, 622=S/L, 65C=Br/C, 65V=Br/Sc, 652=Br/L, 742=Ca/L.

Abbreviations: Ca=Carmine, Br=Bronze, C=Cloth, Gr=Graphite, L=Leather, Mb=Medium Blue, Mg=Medium Gray, S=Saddle, Sc=Sport Seat Cloth.

1986 Corvette

Production: 27,794 coupe, 7,315 convertible, 35,109 total

1986 Numbers

Vehicle: 1G1YY0789G5100001 thru 1G1YY0789G5127794 (coupe)
1G1YY6789G5900001 thru 1G1YY6789G5907315 (conv)
  • Ninth digit is a security code and varies.

Suffix: DKF: 350ci, 230hp, ih, at, oc   ZJS: 350ci, 235hp, ah, at, oc
DKC: 350ci, 230hp, ih, at   ZJH: 350ci, 235hp, ah, at
DKH: 350ci, 230hp, ih, at, ex   ZKD: 350ci, 235hp, ah, at, ex
DKD: 350ci, 230hp, ih, mt, oc   ZJW: 350ci, 235hp, ah, mt, oc
DKB: 350ci, 230hp, ih, mt   ZKA: 350ci, 235hp, ah, mt

Block: 14088548: All

Head: 462624: 350ci, 230hp, ih   14101128: 350ci, 235hp, ah

Abbreviations: ah=aluminum heads, at=automatic transmission, ci=cubic inch, ex=export, hp=horsepower, ih=iron heads, mt=manual transmission, oc=engine oil cooler.

1986 Facts

• A convertible model was introduced, the first Chevrolet-built Corvette convertible since the 1975 model. The 1986 Corvette was the pace car for the 1986 Indianapolis 500. All 1986 Corvette convertibles sold were designated as pace car replicas regardless of color or option content, and all included decal packages for dealer or customer installation. The actual Indy Pace Car was yellow and 732 yellow 1986 convertibles were sold.

• An anti-lock brake system (ABS) became standard with 1986 Corvettes. An adaptation of Bosch's system, Corvette's ABS had rotational sensors at each wheel to transfer data to a computerized electronic control unit (ECU). Brake line pressure was automatically distributed for optimum braking without wheel lock and loss of steering control.

• Cracking around the head attachment bosses required an introduction delay for design revisions to 1986's aluminum cylinder heads. The heads were ready in time for convertible production and all 1986 convertibles and late production coupes had aluminum heads (it is believed the last 8,594 coupes built had aluminum heads). Engines fitted with aluminum heads were rated at 235hp, an increase of 5hp.

• Center high mount stoplights were added to 1986 Corvettes to conform to federal requirements. The coupe's was mounted above the rear window; the convertible's was in a less-conspicuous rear fascia location.

• A new vehicle antitheft system (VATS) required a special ignition key with an embedded pellet. Lock cylinder contacts measured the pellet's electrical resistance (there were fifteen variations) before allowing start.

• Caster was changed in 1986 from 3.8 degrees to 6.0 degrees to improve on-center road feel and to decrease wander.

• Fifty "Malcolm Konner Commemorative Edition" 1986 Corvettes were built in a special arrangement honoring the New Jersey Chevrolet dealership's founder, Malcolm Konner. Each Corvette had special two-tone paint schemes, Silver Beige over Black, coded "spec." Window stickers reflected 4001ZA as the RPO, and a $500 cost for MALCOLM KONNER SP. EDIT. PAI. All were coupes, twenty with manual transmissions, thirty with automatics. Twenty had RPO Z51 suspensions, divided equally between manual and automatic transmissions. All had graphite leather interiors. One became the first Callaway twin-turbo engine conversion. Konner Corvettes were not built in sequence as serial numbers ranged from 1G1YY0781G5108043 to 1G1YY0782G5108343.

• The angle of the 1986 digital instrument display was changed to improve daytime viewing by reducing glare.

• A new upshift indicator light for manual and automatic transmission 1986 models was intended to improve fuel economy.

• "Low coolant" and "anti-lock" instrument displays were added.

• Wheel design was revised slightly for 1986, with the wheel center section natural finish instead of black as in 1984 and 1985.

1986 Options

RPO#	DESCRIPTION	QTY	RETAIL $
1YY07	Base Corvette Sport Coupe	27,794	$27,027.00
1YY67	Base Corvette Convertible	7,315	32,032.00
AG9	Power Driver Seat	33,983	225.00
AQ9	Sport Seats, leather	13,372	1,025.00
AR9	Base Seats, leather	—	400.00
AU3	Power Door Locks	34,215	175.00
B4P	Radiator Boost Fan	8,216	75.00
B4Z	Custom Feature Package	4,832	195.00
C2L	Dual Removable Roof Panels (coupe)	6,242	895.00
24S	Removable Roof Panel, blue tint (coupe)	12,021	595.00
64S	Removable Roof Panel, bronze tint (coupe)	7,819	595.00
C68	Electronic Air Conditioning Control	16,646	150.00
D84	Two-Tone Paint (coupe)	3,897	428.00
FG3	Delco-Bilstein Shock Absorbers	5,521	189.00
G92	Performance Axle Ratio, 3.07:1	4,879	22.00
KC4	Engine Oil Cooler	7,394	110.00
K34	Cruise Control	34,197	185.00
MM4	4-Speed Manual Transmission	6,835	0.00
NN5	California Emission Requirements	5,697	99.00
UL5	Radio Delete	166	-256.00
UM6	AM-FM Stereo Cassette	2,039	122.00
UU8	Stereo System, Delco-Bose	32,478	895.00
V01	Heavy-Duty Radiator	10,423	40.00
Z51	Performance Handling Package (coupe)	12,821	470.00
Z6A	Rear Window+Side Mirror Defog (coupe)	21,837	165.00
4001ZA	Malcolm Konner Special Edition (coupe)	50	500.00

• A 350ci, 230hp (iron cylinder heads) or 235hp (aluminum cylinder heads) engine, 4-speed automatic transmission, removable body-color roof panel (coupe) or soft top (conv), and cloth seats were included in the base price.

• RPO B4Z custom feature package included rear window defogger, outside remote heated mirrors, and inside rearview mirror with map light.

• RPO Z51 included B4P, FG3, KC4, V01, 16x9.5-inch wheels, heavy-duty suspension, and fast steering ratio. Limited to coupes.

• Non-USA sales included 1,670 Canada, 20 Japan, 32 Europe, and 4 other.

1986 Colors

CODE	EXTERIOR	QTY	SOFT TOP	INTERIORS
13	Silver Metallic	1,209	Bk-W	Gr-Mg
18	Medium Gray Metallic	1,603	Bk-W	Gr-Mg-R
20	Medium Blue Metallic	128	Bk-W	B-Gr
35	Yellow	1,464	Bk-W	Gr
40	White	4,176	Bk-S-W	B-Br-Gr-Mg-R-S
41	Black	5,464	Bk-S-W	Gr-Mg-R-S
53	Gold Metallic	777	Bk-S	Gr-S
59	Silver Beige Metallic	1,383	Bk	Br-Gr
66	Copper Metallic	4	Bk-S	Gr-S
69	Medium Brown Metallic	488	S	Br-S
74	Dark Red Metallic	5,002	Bk-S-W	Gr-S
81	Bright Red	9,466	Bk-S-W	Gr-R-S
13/18	Silver/Gray	1,049	none	Gr-Mg-R
18/41	Gray/Black	1,138	none	Gr-Mg
40/13	White/Silver	693	none	Gr-Mg
59/69	Silver Beige/Medium Brown	1,014	none	Br
spec	Silver Beige/Black	50	none	Gr

• Suggested interiors shown. Other combinations were possible.

• Restrictions applied to some soft top and interior color combinations.

• The code "spec" was used for fifty "Malcolm Konner Commemorative Edition" Corvettes with unique Silver Beige over Black two-tone paint. Chevrolet records indicate 51 Corvettes were coded "spec" so it is possible that either one additional car was painted in the Konner colors, or that one car was painted an unlisted color or left in primer.

• The four Copper Metallic (code 66) Corvettes were all coupes.

• Convertible top colors for 1986 were 3,685 black, 2,682 white, 948 saddle.

Interior Codes: 12C=Gr/C, 122=Gr/L, 15C=Mg/C, 152=Mg/L, 21C=B/C, 212=B/L, 62C=S/C, 622=S/L, 65C=Br/C, 652=Br/L, 732=R/L.

Abbreviations: B=Blue, Bk=Black, Br=Bronze, C=Cloth, Gr=Graphite, L=Leather, Mg=Medium Gray, R=Red, S=Saddle, W=White.

1987 Corvette

Production: 20,007 coupe, 10,625 convertible, 30,632 total

1987 Numbers

Vehicle: 1G1YY2182H5100001 thru 1G1YY2182H5130632
  - For convertibles, sixth digit is a 3.
  - Ninth digit is a security code and varies.

Suffix: ZJN: 350ci, 240hp, at   ZLA: 350ci, 240hp, at, oc
ZLC: 350ci, 240hp, mt, oc   ZLB: Export

• For Callaway twin-turbo, Chevrolet engine coding was replaced as follows: First two digits for year, next three digits for Callaway sequence, last four digits to match last four digits of vehicle identification number.

Block: 14093638: All

Head: 14101128: All

Abbreviations: at=automatic transmission, ci=cubic inch, hp=horsepower, mt=manual transmission, oc=engine oil cooler.

1987 Facts

• Friction reduction from roller valve lifters (new for 1987) resulted in a power increase to 240hp, up 5hp from 1986's aluminum-head engines.

• Center sections and radial slots of 1987 wheels were painted argent gray. The radial slots of 1986 wheel were painted, but the wheel centers were not. Centers and radial slots for 1984 and 1985 were painted black.

• Convertibles and early coupes had outside mirror air deflectors.

• Chevrolet planned an RPO UJ6 low tire pressure indicator option, but the $325 option was on constraint during 1987 due to false signalling problems (the signal from one car could set off the alarm of another in close proxmity). However, records show forty-six units installed on 1987 models. The option returned for the 1989 model year.

• The RPO Z51 package was refined to include convertible-derived structural enhancement forward of the dash, and a finned power steering fluid cooler. It was restricted to manual-transmission coupes.

• A new RPO Z52 "sport" handling package combined elements of Z51 with the softer suspension of base models. RPO Z52 included the radiator boost fan, Bilstein shock absorbers, engine oil cooler, heavy-duty radiator, 16x9.5-inch wheels, faster 13:1 steering ratio, larger front stabilizer bar (except early production), and the convertible-inspired structural improvements for coupes. Z52 was available with coupes or convertibles, automatic or manual transmissions.

• The overdrive-engaged light was moved from the center-dash area (1984-86) to an easier-to-view location within the 1987 tach display.

• The Callaway Twin-Turbo engine package introduced in 1987 was not a factory-installed option, but could be ordered through participating Chevrolet dealers as RPO B2K. Fully assembled Corvettes were shipped from the Bowling Green Corvette plant to Callaway Engineering in Old Lyme, Connecticut, for engine and other modifications. The 1987 Callaway had ratings of 345hp and 465 lb.-ft torque, and reached a top speed of 177.9 mph with .60 overdrive gearing. The first four 1987 Callaways used replacement LF5 (truck) shortblocks, but subsequent cars had reworked production Corvette engines. All 1987 Callaways had manual transmissions and none were certified for California sale. Of 188 twin-turbos built in 1987, 123 were coupes, 65 were convertibles.

• Electronic air conditioning control (RPO C68) became an available option for coupes and convertibles in 1987; in 1986, it was coupe-only.

• New 1987 convenience options included an illuminated vanity mirror (D74) for the driver's visor, and a passenger-side power seat base (AC1). Twin remote heated mirrors became available for convertibles as RPO DL8. The heated mirrors were included with the heated rear window in the RPO Z6A defogger option for coupes.

• In its January 1987 "top ten" issue, *Car and Driver* magazine picked the 1987 Corvette as one of the world's top ten cars. It also won the 70-0mph braking award in the top ten performer's category. In a *Motor Trend* magazine "Domestic Dynamite" article (August 1987), Corvette won in top speed, best braking and best skidpad categories.

1987 Options

RPO #	DESCRIPTION	QTY	RETAIL $
1YY07	Base Corvette Sport Coupe	20,007	$27,999.00
1YY67	Base Corvette Convertible	10,625	33,172.00
AC1	Power Passenger Seat	17,123	240.00
AC3	Power Driver Seat	29,561	240.00
AQ9	Sport Seats, leather	14,119	1,025.00
AR9	Base Seats, leather	14,561	400.00
AU3	Power Door Locks	29,748	190.00
B2K	Callaway Twin Turbo (not GM installed)	188	19,995.00
B4P	Radiator Boost Fan	7,291	75.00
C2L	Dual Removable Roof Panels (coupe)	5,017	915.00
24S	Removable Roof Panel, blue tint (coupe)	8,883	615.00
64S	Removable Roof Panel, bronze tint (coupe)	5,766	615.00
C68	Electronic Air Conditioning Control	20,875	150.00
DL8	Twin Remote Heated Mirrors (convertible)	6,840	35.00
D74	Illuminated Driver Vanity Mirror	14,992	58.00
D84	Two-Tone Paint (coupe)	1,361	428.00
FG3	Delco-Bilstein Shock Absorbers	1,957	189.00
G92	Performance Axle Ratio, 3.07:1	7,286	22.00
KC4	Engine Oil Cooler	6,679	110.00
K34	Cruise Control	29,594	185.00
MM4	4-Speed Manual Transmission	4,298	0.00
NN5	California Emission Requirements	5,423	99.00
UL5	Radio Delete	247	-256.00
UM6	AM-FM Stereo Cassette	2,236	132.00
UU8	Stereo System, Delco-Bose	27,721	905.00
V01	Heavy-Duty Radiator	7,871	40.00
Z51	Performance Handling Package (coupe)	1,596	795.00
Z52	Sport Handling Package	12,662	470.00
Z6A	Rear Window+Side Mirror Defog (coupe)	19,043	165.00

• A 350ci, 240hp engine, 4-speed automatic transmission, removable body-color roof panel (coupe) or soft top (convertible), and cloth seats were included in the base price.

• RPO Z51 included B4P, FG3, KC4, V01, 16x9.5-inch wheels, heavy-duty suspension, fast steering ratio. Coupes and manual transmissions only.

• RPO B2K (Callaway Twin Turbo) generated a specific equipment build with standard engines at the Corvette assembly plant. The cars were then drop-shipped to Callaway's Connecticut shop for installation of Callaway-modified twin-turbo engines.

• Non-USA sales included 1,566 Canada, 25 Japan, 13 Europe, 3 Gulf States including Saudi Arabia. Speedometers with 105 kh (kilometer per hour) warning total was 14.

1987 Colors

CODE	EXTERIOR	QTY	SOFT TOP	INTERIORS
13	Silver Metallic	767	Bk-W	Gr-Mg
18	Medium Gray Metallic	1,035	Bk-W	Gr-Mg-R
20	Medium Blue Metallic	2,677	Bk-W	B-Gr
35	Yellow	1,051	Bk-W	Gr
40	White	3,097	Bk-S-W	B-Br-Gr-Mg-R-S
41	Black	5,101	Bk-S-W	Gr-Mg-R-S
53	Gold Metallic	397	Bk-S	Gr-S
59	Silver Beige Metallic	950	Bk	Br-Gr
66	Copper Metallic	87	Bk-S	Gr-S
69	Medium Brown Metallic	245	S	Br-S
74	Dark Red Metallic	5,578	Bk-S-W	Gr-S
81	Bright Red	8,285	Bk-S-W	Gr-R-S
13/18	Silver/Gray	403	none	Gr-Mg-R
18/41	Gray/Black	316	none	Gr-Mg
40/13	White/Silver	195	none	Gr-Mg
59/69	Silver Beige/Medium Brown	447	none	Br

• Suggested interiors shown. Other combinations were possible.

• Restrictions applied to some soft top and interior color combinations.

• Convertible top colors for 1987 were 4,748 black, 4,158 white, 1,719 saddle.

Interior Codes: 12C=Gr/C, 122=Gr/L, 15C=Mg/C, 152=Mg/L, 21C=B/C, 212=B/L, 62C=S/C, 622=S/L, 65C=Br/C, 652=Br/L, 732=R/L.

Abbreviations: B=Blue, Bk=Black, Br=Bronze, C=Cloth, Gr=Graphite, L=Leather, Mg=Medium Gray, R=Red, S=Saddle, W=White.

1988 Corvette

1988 Numbers

Vehicle: 1G1YY2182J5100001 thru 1G1YY2182J5122789
- For convertibles, sixth digit is a 3.
- Ninth digit is a security code and varies.

Suffix: ZMA: 350ci, 240/245hp, at   ZMC: 350ci, 240/245hp, mt, oc
ZMD: 350ci, 240/245hp, at, oc

• For Callaway twin-turbo, Chevrolet engine coding was replaced as follows: First two digits for year, next three digits for Callaway sequence, last four digits to match last four digits of vehicle identification number.

Block: 14093638: All

Head: 10088113: All

Abbreviations: at=automatic transmission, ci=cubic inch, hp=horsepower, mt=manual transmission, oc=engine oil cooler.

1988 Facts

• Refinements for 1988 included carpeted door sills, solution-dyed carpet, improved "flow through" ventilation for coupes, and a lower, rearward relocation of the parking brake handle.

• Engine power remained at 240hp for 1988 models except for coupes with 3.07:1 axle ratios which had 245hp. The 5hp increase came from less restrictive mufflers which were deemed too loud for convertibles and 2.59:1 axle coupes.

• A thirty-fifth anniversary edition 1988 Corvette package was available for coupes only. It featured a two-tone exterior of white with black roof bow, white leather seats and steering wheel, special interior and exterior accents, a console-mounted anniversary plaque, special emblems and other distinguishing features. Sales totaled 2,050 units.

• Chevrolet built fifty-six street-legal Corvettes for the 1988 SCCA Corvette Challenge race series. Engines, stock but matched for power output, were built at the Flint engine plant, sealed and shipped to Bowling Green for standard assembly. The cars weren't built in sequence because the Corvette plant built in color batches. Fifty cars were sent to Protofab in Wixom, Michigan for installation of roll cages and other gear. During the season, most engines were exchanged by Chevrolet for new, sealed engines with more evenly calibrated power output.

• New six-slot 16x8.5 wheels were standard with P255/50ZR16 tires.

• RPO Z51 and RPO Z52 content changed slightly for 1988. Both had newly styled 17x9.5-inch wheels with twelve cooling slots, and P275/40ZR17 tires. Z51 had higher spring rates and finned power steering cooler as before, but in 1988 it also received larger front brake rotors and calipers. RPO Z51 was limited to manual transmission coupes. RPO Z52 was not restricted.

• All 1988 Corvettes had new dual-piston front brakes and parking brakes which activated the rear pads instead of activating small, separate parking drum brakes as in all previous disc-brake Corvettes.

• Ratings for the 1988 RPO B2K Callaway Twin Turbo were 382hp and 562 lb.-ft torque. Automatic transmissions (reworked truck Turbo Hydra-Matic 400s) were available for $6,500. Either Z51 or Z52 suspensions could be specified. Later production had Z52 then Z51's larger front brakes, less restrictive mufflers, longer air dams, and steering coolers because RPO B2K triggered these through Special Equipment Option (SEO) Z5G. Engines were reworked by Callaway at its Old Lyme, Connecticut shop. Full Chevrolet warranty applied except for powertrain which was covered by Callaway for twelve months or 12,000 miles.

• All 1988s had an improved hood support rod and a more efficient air conditioning compressor manufactured by Nippondenso. Nippondenso also supplied a new gear-reduction starter motor.

• Front suspensions were redesigned for zero scrub radius to reduce steering-wheel fight from brake torque and road inputs. This included revised steering knuckles, upper/lower control arms, springs, shock absorbers, stabilizer shaft/link assemblies, and wheel bearing assemblies.

1988 Options

RPO#	DESCRIPTION	QTY	RETAIL $
1YY07	Base Corvette Sport Coupe	15,382	$29,489.00
1YY67	Base Corvette Convertible	7,407	34,820.00
AC1	Power Passenger Seat	18,779	240.00
AC3	Power Driver Seat	22,084	240.00
AQ9	Sport Seats, leather	12,724	1,025.00
AR9	Base Seats, leather	9,043	400.00
B2K	Callaway Twin Turbo (not GM installed)	125	25,895.00
B4P	Radiator Boost Fan	19,035	75.00
C2L	Dual Removable Roof Panels (coupe)	5,091	915.00
24S	Removable Roof Panel, blue tint (coupe)	8,332	615.00
64S	Removable Roof Panel, bronze tint (coupe)	3,337	615.00
C68	Electronic Air Conditioning Control	19,372	150.00
DL8	Twin Remote Heated Mirrors (convertible)	6,582	35.00
D74	Illuminated Driver Vanity Mirror	14,249	58.00
FG3	Delco-Bilstein Shock Absorbers	18,437	189.00
G92	Performance Axle Ratio, 3.07:1	4,497	22.00
KC4	Engine Oil Cooler	18,877	110.00
MM4	4-Speed Manual Transmission	4,282	0.00
NN5	California Emission Requirements	3,882	99.00
UL5	Radio Delete	179	-297.00
UU8	Stereo System, Delco-Bose	20,304	773.00
V01	Heavy-Duty Radiator	19,271	40.00
Z01	35th Special Edition Package (coupe)	2,050	4,795.00
Z51	Performance Handling Package (coupe)	1,309	1,295.00
Z52	Sport Handling Package	16,017	970.00
Z6A	Rear Window+Side Mirror Defog (coupe)	14,648	165.00

• A 350ci, 240/245hp engine, 4-speed automatic transmission, removable body-color roof panel (coupe) or soft top (convertible), and cloth seats were included in the base price.

• RPO Z51 included B4P, FG3, KC4, V01, heavy-duty suspension, 17x9.5 wheels, and fast steering ratio. Limited to manual transmission coupes.

• RPO Z52 included B4P, FG3 and KC4, 17x9.5 wheels, and fast steering.

• RPO Z01 included AQ9, AC3, 24S, C68, D74, Z52 and Z6A. Not available with convertibles.

• RPO B2K generated a specific equipment build with standard engines at the Corvette assembly plant. The cars were then sent to Callaway's Connecticut shop for installation of Callaway-modified twin-turbo engines.

• Non USA sales included 1,305 Canada, 59 Japan. and 108 Europe.

1988 Colors

CODE	EXTERIOR	QTY	SOFT TOP	INTERIORS
13	Silver Metallic	385	Bk-W	Bk-G-R-S
20	Medium Blue Metallic	1,148	Bk-W	B-Bk
28	Dark Blue Metallic	1,675	Bk-S-W	Bk-S
35	Yellow	578	Bk-W	Bk-S
40	White	3,620	Bk-S-W	B-Bk-G-R-S
41	Black	3,420	Bk-S-W	Bk-G-R-S
74	Dark Red Metallic	2,878	Bk-S-W	Bk-S
81	Bright Red	5,340	Bk-S-W	Bk-G-R-S
90	Gray Metallic	644	Bk-W	Bk-G
96	Charcoal Metallic	1,046	Bk-S-W	G-S
40/41	White/Black	2,050	none	W

• Suggested interiors shown. Other combinations were possible. Some soft top and interior color combinations were restricted.

• Two-tones of gray/charcoal and white/silver were indicated in early dealer order guides, but were dropped prior to 1988 production.

• Five code 66 dark orange exteriors were built for pilot production.

• Convertible top colors for 1988 were 3,431 black, 2,811 white, 1,165 saddle.

• Interior colors sold in 1988 were 2,050 white sport leather, 3,747 black sport leather, 919 blue sport leather, 1,757 saddle sport leather, 2,913 red sport leather, 1,338 gray sport leather, 3,087 black base leather, 789 blue base leather, 1,449 saddle base leather, 2,492 red base leather, 1,226 gray base leather, 736 black cloth and 286 saddle cloth.

Interior Codes: 113=W/L, 19C=Bk/C, 192=Bk/L, 212=B/L, 60C=S/C, 602=S/L, 732=R/L, 902=G/L.

Abbreviations: B=Blue, Bk=Black, C=Cloth, G-Gray, L=Leather, R=Red, S=Saddle, W=White.

1989 Corvette

Production: 16,663 coupe, 9,749 convertible, 26,412 total

1989 Numbers

Vehicle: 1G1YY2186K5100001 thru 1G1YY2186K5126328
- For convertibles, sixth digit is a 3.
- Ninth digit is a security code and varies.

Suffix: ZRA: 350ci, 240/245hp, mt, oc   ZRC: 350ci, 240/245hp, at, oc
ZRB: 350ci, 240/245hp, at

• Production totals included 84 ZR-1s built, but not sold to the public.

• For Callaway twin-turbo, Chevrolet engine coding was replaced as follows: First two digits for year, next three digits for Callaway sequence, last four digits to match last four digits of vehicle identification number.

Block: 14093638: All

Head: 10088113: All

Abbreviations: at=automatic transmission, ci=cubic inch, hp=horsepower, mt=manual transmission, oc=engine oil cooler.

1989 Facts

• The RPO MN6 manual transmission no-cost option for 1989 was a new 6-speed unit designed jointly by ZF (Zahnradfabrik Friedrichshafen) and Chevrolet, and initially built by ZF in Germany. A computer-aided gear selection (CAGS) feature bypassed second and third gears (and locked out fifth and sixth) for improved fuel economy when a series of low-performance criteria were met.

• The Corvette Challenge race series terminated at the end of the 1989 season. For 1989, Corvette's Bowling Green plant built sixty Challenge cars with standard engines. Meanwhile, CPC Flint Engine built special, higher horsepower engines which were shipped to the Milford Proving Grounds for storage, then to Specialized Vehicles Inc. (SVI), Troy, Michigan, where they were equalized for power and sealed. Bowling Green sent thirty cars to Powell Development America, Wixom, Michigan, where the roll cages and safety equipment were installed and the engines from SVI were switched with the original engines. At season's end, Chevrolet returned the original numbers-matching engines to each racer.

• The RPO Z51 performance handing package option continued in 1989, available only in coupes with manual transmissions. A new suspension option, RPO FX3, permitted three variations of suspension control regulated by a console switch. It could be ordered only with RPO Z51. The RPO Z52 sport suspension (1987-1988) was not a 1989 option. However, though all 1989 Corvettes with FX3 were Z51s, these had Z52 springs and stabilizers for a wider range of suspension control. The only exceptions were the sixty Corvettes built for the Challenge race series which had FX3 suspensions with Z51 springs and stabilizers.

• The standard six-slot, 16x8.5-inch wheel introduced in 1988 was discontinued for 1989. The twelve-slot, 17x9.5-inch style included with 1988's Z51 and Z52 options became 1989's standard equipment wheel.

• The RPO UJ6 low-tire pressure warning system, originally announced for 1987 models, was available for 1989 models. Sensors strapped to the inside of each wheel sent a radio signal to a instrument-panel receiver if pressure in any tire dropped to a preset limit.

• After a tremendous media blitz, Chevrolet advised dealers on April 19, 1989 that the ZR-1 would be a 1990 model, not a late-release 1989. The reason cited was "insufficient availability of engines caused by additional development." Eighty-four (84) 1989 ZR-1 Corvettes were built for evaluation, testing, media preview and photography, but none were released for public sale.

• Seats were restyled, but the three choices of cloth, optional leather, or optional sport leather continued. Due to weight and fuel economy factors, Chevrolet intentionally limited sales of the sport leather seats by making them available exclusively with Z51-optioned models during 1989.

• The manual top mechanism was simplified for 1989 convertibles.

1989 Options

RPO#	DESCRIPTION	QTY	RETAIL $
1YY07	Base Corvette Sport Coupe	16,663	$31,545.00
1YY67	Base Corvette Convertible	9,749	36,785.00
AC1	Power Passenger Seat	20,578	240.00
AC3	Power Driver Seat	25,606	240.00
AQ9	Sport Seats, leather	1,777	1,025.00
AR9	Base Seats, leather	23,364	400.00
B2K	Callaway Twin-Turbo (not GM installed)	67	25,895.00
B4P	Radiator Boost Fan	20,281	75.00
CC2	Auxiliary Hardtop (convertible)	1,573	1,995.00
C2L	Dual Removable Roof Panels (coupe)	5,274	915.00
24S	Removable Roof Panel, blue tint (coupe)	8,748	615.00
64S	Removable Roof Panel, bronze tint (coupe)	4,042	615.00
C68	Electronic Air Conditioning Control	24,675	150.00
D74	Illuminated Driver Vanity Mirror	17,414	58.00
FX3	Selective Ride and Handling, electronic	1,573	1,695.00
G92	Performance Axle Ratio	10,211	22.00
K05	Engine Block Heater	2,182	20.00
KC4	Engine Oil Cooler	20,162	110.00
MN6	6-Speed Manual Transmission	4,113	0.00
NN5	California Emission Requirements	4,501	100.00
UJ6	Low Tire Pressure Warning Indicator	6,976	325.00
UU8	Stereo System, Delco-Bose	24,145	773.00
V01	Heavy-Duty Radiator	20,888	40.00
V56	Luggage Rack (convertible)	616	140.00
Z51	Performance Handling Package (coupe)	2,224	575.00

• A 350ci, 240/245hp engine, 4-speed automatic transmission, removable body-color roof panel (coupe) or soft top (convertible), and cloth seats were included in the base price.

• RPO Z51 included B4P,KC4, V01, heavy-duty suspension, and fast steering ratio. Limited to manual transmission coupes.

• New convertible options for late 1989 introduction included an RPO V56 rear luggage rack, and RPO CC2 removable hardtop.

• RPO AQ9 leather sport seats and RPO FX3 selective ride and handling were available only when ordered with RPO Z51.

• RPO B2K generated a specific equipment build with standard engines at the Corvette assembly plant. The cars were then sent to Callaway's Connecticut shop for installation of Callaway-modified twin-turbo engines.

• Chevrolet records show 1 radio delete Corvette coupe built in 1989.

• Non USA sales in 1989 included 1,174 Canada, 110 Japan, 201 Europe.

1989 Colors

CODE	EXTERIOR	QTY	SOFT TOP	INTERIORS
10	White	5,426	Bk-S-W	B-Bk-G-R-S
20	Medium Blue Metallic	1,428	Bk-W	B-Bk
28	Dark Blue Metallic	1,931	Bk-S-W	Bk-S
41	Black	4,855	Bk-S-W	B-Bk-G-R-S
68	Dark Red Metallic	3,409	Bk-S-W	Bk-S
81	Bright Red	7,663	Bk-S-W	Bk-G-R-S
90	Gray Metallic	225	Bk-W	Bk-G
96	Charcoal Metallic	1,440	Bk-S-W	Bk-G-S

• Only interior-exterior combinations shown were considered acceptable.

• Restrictions applied to some soft top and interior color combinations.

• Code 90 Gray was cancelled in November, 1988.

• Though not listed on 1989 exterior color availability charts, Chevrolet records indicate 8 code 35 Yellow/Buckskin (all coupes), and 27 code 31 Arctic Pearl 1989 Corvettes (24 coupes, 3 convertibles) were built.

• Interior colors sold in 1989 were 871 black base cloth, 400 saddle base cloth, 8,186 black base leather, 4,159 saddle base leather, 1,428 blue base leather, 6,714 red base leather, 2,877 gray base leather, 852 black sport leather, 226 saddle sport leather, 55 blue sport leather, 425 red sport leather, 219 gray sport leather.

Interior Codes: 19C=Bk/C, 192=Bk/L, 212=B/L, 60C=S/C, 602=S/L, 732=R/L, 902=G/L.

Abbreviations: B=Blue, Bk=Black, C=Cloth, G=Gray, L=Leather, R=Red, S=Saddle, W=White.

1990 Corvette

Production: 16,016 coupe, 7,630 convertible, 23,646 total

1990 Numbers

Vehicle: 1G1YY2380L5100001 thru 1G1YY2380L5120597
1G1YZ23J6L5800001 thru 1G1YZ23J6L5803049 (ZR1)
    • For convertibles, sixth digit is a 3.
    • Ninth digit is a security code and varies.

Suffix: ZSA: 350ci, 245/250hp, at   ZSD: 350ci, 375hp, ac
           ZSB: 350ci, 245/250hp, mt, oc   ZSH: 350ci, 375hp, ea
           ZSC: 350ci, 245/250hp, at, oc

• For Callaway twin-turbo, Chevrolet engine coding was replaced as follows: First two digits for year, next three digits for Callaway sequence, last four digits to match last four digits of vehicle identification number.

Block: 10090511: 350ci, 375hp   14093638: 350ci, 245/250hp

Head: 10088113: 350ci, 245/250hp
      10106178: 350ci, 375hp (rh)   10106179: 350ci, 375hp (lh)

Abbreviations: ac=air conditioning, at=automatic transmission, ci=cubic inch, ea=electronic air conditioning control, hp=horsepower, lh=left hand, mt=manual transmission, oc=engine oil cooler, rh=right hand.

1990 Facts

• The ZR-1 (RPO ZR1) arrived as a 1990 model after much anticipation. Its LT5 engine was designed with the same V-8 configuration and 4.4-inch bore spacing as the standard L98 Corvette engine, but was an otherwise new Lotus-Chevrolet design with four overhead camshafts and 32 valves. Despite extensive use of aluminum, the complexity of the LT5 engine caused it to weigh just under forty pounds more than the base iron-block Corvette engine. The LT5 produced 375hp, yet avoided the EPA's gas guzzler tax. LT5 engines were manufactured and assembled by Mercury Marine in Stillwater, Oklahoma, then shipped to the Corvette Bowling Green assembly plant for ZR-1 vehicle assembly.

• For a limited time, dealers could order Corvettes destined for the World Challenge race series. Merchandising code R9G triggered deviations from normal build, such as heavy-duty springs with FX3. Owners could buy sealed race engines from Chevrolet; modifications were the owner's responsibility. Twenty-three 1990 R9G Corvettes were built.

• An air intake speed density control system, camshaft revision, and compression ratio increase added 5hp to base-engines, up from 240hp to 245hp (except coupes with 3.07:1 or 3.33:1 axle ratios which increased from 245hp to 250hp because of their less-restrictive exhaust systems).

• 1990 Corvettes had improved ABS and improved yaw control.

• An engine oil life monitor calculated useful oil life based on engine temperatures and revolutions. An instrument panel display alerted the driver when an oil change was recommended.

• The RPO V01 radiator and B4P boost fan were not optional in 1990, both unnecessary due to 1990's more efficient, sloped radiator design.

• Two premium 200-watt Delco-Bose stereo systems were available, the top unit featuring a compact disc player. To discourage theft, the CD required electronic security code input after battery disconnect.

• The instrument panel for 1990 was redesigned as a "hybrid," combining a digital speedometer with analog tachometer and secondary gauges. A supplemental inflatable restraint system (SIR) with airbag was added to the driver side, and a glovebox to the passenger side.

• The "ABS Active" light was removed from the driver information center.

• Seat designs were the same for 1990 as the previous year, except the backs would latch in the forward position. Sport seats were available with all models and suspensions in 1990.

• Chevrolet service departments returned LT5 engines to Mercury Marine for certain repairs. Customers had the choice of a replacement engine, or return of their original engine if repairable.

• Corvette did not pace the 1990 Indy 500, but 80 turquoise or yellow "Indy Festival" convertibles with special graphics were built for the event.

1990 Options

RPO #	DESCRIPTION	QTY	RETAIL $
1YY07	Base Corvette Sport Coupe	16,016	$31,979.00
1YY67	Base Corvette Convertible	7,630	37,264.00
AC1	Power Passenger Seat	20,419	270.00
AC3	Power Driver Seat	23,109	270.00
AQ9	Sport Seats, leather	11,457	1,050.00
AR9	Base Seats, leather	11,649	425.00
B2K	Callaway Twin-Turbo (not GM installed)	58	26,895.00
CC2	Auxiliary Hardtop (convertible)	2,371	1,995.00
C2L	Dual Removable Roof Panels (coupe)	6,422	915.00
24S	Removable Roof Panel, blue tint (coupe)	7,852	615.00
64S	Removable Roof Panel, bronze tint (coupe)	4,340	615.00
C68	Electronic Air Conditioning Control	22,497	180.00
FX3	Selective Ride and Handling, electronic	7,576	1,695.00
G92	Performance Axle Ratio	9,362	22.00
K05	Engine Block Heater	1,585	20.00
KC4	Engine Oil Cooler	16,221	110.00
MN6	6-Speed Manual Transmission	8,101	0.00
NN5	California Emission Requirements	4,035	100.00
UJ6	Low Tire Pressure Warning Indicator	8,432	325.00
UU8	Stereo System, Delco-Bose	6,701	823.00
U1F	Stereo System with CD, Delco-Bose	15,716	1,219.00
V56	Luggage Rack (convertible)	1,284	140.00
Z51	Performance Handling Package (coupe)	5,446	460.00
ZR1	Special Performance Package (coupe)	3,049	27,016.00

• A 350ci, 245/250hp engine, 4-speed automatic transmission, removable body-color roof panel (coupe) or soft top (convertible), and cloth seats were included in the base price.

• RPO Z51 included KC4, heavy-duty suspension and brakes. Available with coupe and manual transmission only.

• RPO ZR1 included unique bodywork (doors, rear quarters, rockers, rear fascia, and rear upper panel) to accept Goodyear Z-rated P315/35ZR17 tires on 11-inch wide rear rims. RPOs AC1, AC3, AQ9, FX3, LT5 (32-valve engine, exclusive to the ZR-1), U1F, UJ6 and a specially laminated "solar" windshield were included. RPO MN6 manual transmission was required. Available in coupe body style only.

• RPO K05 engine block heater was not available with RPO ZR1.

• Non USA sales were 951 Canada, 176 Japan, 951 Europe.

1990 Colors

CODE	EXTERIOR	QTY	SOFT TOP	INTERIORS
10	White	4,872	Bk-S-W	B-Bk-G-R-S
25	Steel Blue Metallic	813	Bk-W	B-Bk
41	Black	4,759	Bk-W	B-Bk-G-R
42	Turquoise Metallic	589	Bk-S	Bk-S
53	Competition Yellow	278	Bk-S-W	Bk-G-S
68	Dark Red Metallic	2,353	Bk-S-W	Bk-S
80	Quasar Blue Metallic	474	Bk-S	Bk-S
81	Bright Red	6,956	Bk-S-W	Bk-G-R-S
91	Polo Green Metallic	1,674	Bk-S	Bk-S
96	Charcoal Metallic	878	Bk-S	Bk-G

• Only interior-exterior combinations shown were considered acceptable.

• Restrictions applied to some soft top and interior color combinations.

• Codes 42, 53 and 80 were not available early.

• Code 53 Competition Yellow exterior was discontinued 5-11-90 due to pigment photosensitivity which caused the paint to temporarily darken after sunlight exposure.

• Convertible top colors for 1990 were 4,161 black, 2,205 white, 1,284 saddle.

• Interior colors sold in 1990 were 398 black base cloth, 142 saddle base cloth, 4,701 black base leather, 395 blue base leather, 2,050 saddle base leather, 3,070 red base leather, 1,433 gray base leather, 4,977 black sport leather, 341 blue sport leather, 1,952 saddle sport leather, 2,818 red sport leather, 1,369 gray sport leather.

Interior Codes: 19C=Bk/C, 193=Bk/L, 223=B/L, 60C=S/C, 603=S/L, 733=R/L, 903=G/L.

Abbreviations: B=Blue, Bk=Black, C=Cloth, G=Gray, L=Leather, R=Red, S=Saddle, W=White.

1991 Corvette

Production: 14,967coupe, 5,672 convertible, 20,639 total

1991 Numbers

Vehicle: 1G1YY2386M5100001 thru 1G1YY2386M5118595
1G1YZ23J6M5800001 thru 1G1YZ23J6M5802044 (ZR-1)
 • For convertibles, sixth digit is a 3.
 • Ninth digit is a security code and varies.

Suffix: ZTA: 350ci, 245/250hp, at   ZTC: 350ci, 245/250hp, at, oc
ZTB: 350ci, 245/250hp, mt, oc   ZTK: 350ci, 375hp

• For Callaway twin-turbo, Chevrolet engine coding was replaced as
follows: First two digits for year, next three digits for Callaway sequence,
last four digits to match last four digits of vehicle identification number.

Block: 10153558: 350ci, 375hp   14093638: 350ci, 245/250hp

Head: 10088113: 350ci, 245/250hp
10174389: 350ci, 375hp (lh)   10174390: 350ci, 375hp (rh)

Abbreviations: at=automatic transmission, ci=cubic inch,
hp=horsepower, lh=left hand, mt=manual transmission, oc=engine oil
cooler, rh=right hand.

1991 Facts

• All 1991 Corvettes had restyled rear exteriors which were similar in
appearance to the 1990 ZR-1 because both had convex rear fascias with
four rectangular tail lamps. Both standard models and ZR-1s also featured
new front designs with wraparound parking-cornering-fog lamps, new
side panel louvers, and wider body-side moldings in body color.

• Despite similar appearance, the 1991 ZR-1 still received unique doors
and rear body panels to accept 11-inch wide rear wheels. The high-mount
center stop lamp for 1991 ZR-1s continued to be roof-mounted. For all
other than ZR-1, the lamp was integrated into the new rear fascia.

• Base wheels were same size as 1990 (17x9.5), but were a new design.

• Finned power steering coolers were included with all 1991 models.

• A new option, RPO Z07, essentially combined the previously available
Z51 performance handling package with FX3 selective ride/handling.
But there were differences. In 1990, if FX3 and Z51 were combined, some
base suspension components were used to provide an adjustable
suspension range from soft to firm. The new RPO Z07 option used all
heavy-duty suspension parts so the ride adjusted from firm to very firm.
Intended for aggressive driving or competition, Z07 was limited to coupes.

• The World Challenge race series continued in 1991, but Bowling Green
did not build specific Corvettes for the series. All race modifications
were the owners' responsibility.

• Callaway Twin-Turbo conversions ended with the 1991 model year.
Callaway built the 500th twin-turbo on 9-26-91 and subsequent builds
were specially badged and optioned (extra $600) as "Callaway 500."

• A power wire for cellular phones or other 12-volt devices was added.

• A power delay feature was added to all models which permitted the
stereo system and power windows to operate after the ignition was
switched to "off" or "lock." Power was cut after the driver door was
opened, or after fifteen minutes, whichever occurred first.

• A sensor utilizing an oil pan float was added to all models. The words
"low oil" appeared in the driver information center for a low oil condition.

• Mufflers were revised for 1991 with larger section sizes and better
control tuning of exhaust note. The mufflers had lower back pressure
for improved performance, but power ratings were not changed.

• The AM band for radios was expanded to receive more frequencies.

• The ZR-1 "valet" power access system continued, but was revised to
default to normal power on each ignition cycle. The full power light was
relocated next to the valet key.

• In a Porsche 911 versus Corvette ZR-1 comparison, *Car and Driver*
magazine gave the nod to Corvette in its April 1991 issue.

1991 Options

RPO #	DESCRIPTION	QTY	RETAIL $
1YY07	Base Corvette Sport Coupe	14,967	$32,455.00
1YY67	Base Corvette Convertible	5,672	38,770.00
AR9	Base Seats, leather	9,505	425.00
AQ9	Sport Seats, leather	10,650	1,050.00
AC1	Power Passenger Seat	17,267	290.00
AC3	Power Driver Seat	19,937	290.00
B2K	Callaway Twin-Turbo (not GM installed)	71	33,000.00
CC2	Auxiliary Hardtop (convertible)	1,230	1,995.00
C2L	Dual Removable Roof Panels (coupe)	5,031	915.00
24S	Removable Roof Panel, blue tint (coupe)	6,991	615.00
64S	Removable Roof Panel, bronze tint (coupe)	3,036	615.00
C68	Electronic Air Conditioning Control	19,233	180.00
FX3	Selective Ride and Handling, electronic	6,894	1,695.00
G92	Performance Axle Ratio	3,453	22.00
KC4	Engine Oil Cooler	7,525	110.00
MN6	6-Speed Manual Transmission	5,875	0.00
NN5	California Emission Requirements	3,050	100.00
UJ6	Low Tire Pressure Warning Indicator	5,175	325.00
UU8	Stereo System, Delco-Bose	3,786	823.00
U1F	Stereo System with CD, Delco-Bose	15,345	1,219.00
V56	Luggage Rack (convertible)	886	140.00
Z07	Adjustable Suspension Package (coupe)	733	2,155.00
ZR1	Special Performance Package (coupe)	2,044	31,683.00

• A 350ci, 245/250hp engine, 4-speed automatic transmission, removable body-color roof panel (coupe) or soft top (convertible), and cloth seats were included in the base price.

• RPO Z07 included KC4, FX3, heavy-duty springs, shocks, stabilizers and bushings, and heavy-duty brakes. Manual transmission was specified, but build records indicate 169 Z07 had automatics.

• RPO ZR1 included unique bodywork (doors, rear quarters, rear fascia, and rear upper panel) to accept Goodyear Z-rated P315/35ZR17 tires on 11-inch wide rear rims. RPOs AC1, AC3, AQ9, C68, FX3, LT5 (32-valve engine exclusive to the ZR-1), U1F, UJ6, and a specially laminated "solar" windshield were included. RPO MN6 manual transmission was required. Available in coupe body style only.

• RPO K05 engine block heater was available with base engines and intended for Canada export only. In 1991, 334 were sold at $20.

• Non USA sales were 484 Canada, 187 Japan, 338 Europe, 1,113 Mexico.

1991 Colors

CODE	EXTERIOR	QTY	SOFT TOP	INTERIORS
10	White	4,305	Bk-Bg-W	B-Bk-G-R-S
25	Steel Blue Metallic	835	B-Bk-W	B-Bk
35	Yellow	650	Bk-Bg-W	Bk-G-S
41	Black	3,909	Bk-Bg-W	B-Bk-G-S-R
42	Turquoise Metallic	1,621	B-Bk	Bk-S
75	Dark Red Metallic	1,311	Bk-Bg-W	Bk-S
80	Quasar Blue Metallic	1,038	Bk-Bg	Bk-S
81	Bright Red	5,318	Bk-Bg-W	Bk-G-R-S
91	Polo Green Metallic	1,230	Bk-Bg	Bk-S
96	Charcoal Metallic	417	Bk-W	Bk-G-S

• Only interior-exterior combinations shown were considered acceptable.

• Restrictions applied to some soft top and interior color combinations.

• Dark Red Metallic (also called Brilliant Red) paint code 75 was the same color as 1990's code 68.

• The blue soft top was not available early in production. It was restricted to code 25 Steel Blue Metallic and code 42 Turquoise Metallic.

• Convertible top colors sold in 1991 were 3,158 black, 1,402 white, 950 beige, 162 blue.

• Interior colors sold in 1991 were 10,376 black, 4,481 red, 3,084 saddle, 1,889 gray, 809 blue (Black cloth was 379; Saddle cloth was 105).

Interior Codes: 19C=Bk/C, 193=Bk/L, 223=B/L, 60C=S/C, 603=S/L, 733=R/L, 903=G/L.

Abbreviations: B=Blue, Bk=Black, Bg=Beige, C=Cloth, G=Gray, L=Leather, R=Red, S=Saddle, W=White.

1992 Corvette

Production: 14,604 coupe, 5,875 convertible, 20,479 total

1992 Numbers

Vehicle: 1G1YY23P6N5100001 thru 1G1YY23P6N5119977
1G1YZ23J6N5800001 thru 1G1YZ23J6N5800502 (ZR-1)
- For convertibles, sixth digit is a 3.
- Ninth digit is a security code and varies.

Suffix: ZAA: 350ci, 375hp   ZUB: 350ci, 300hp, mt
ZAC: 350ci, 300hp, at

Block: 10153558: 350ci, 375hp   10125327: 350ci, 300hp

Head: 10128374: 350ci, 300hp
10174389: 350ci, 375hp (lh) 10174390: 350ci, 375hp (rh)

Abbreviations: at=automatic transmission, ci=cubic inch,
hp=horsepower, lh=left hand, mt=manual transmission, rh=right hand.

1992 Facts

- Exterior appearance for 1992 was little changed. For the ZR-1, "ZR-1" emblems were added above the side fender vents. Two rectangular exhaust outlets were used for ZR-1s and for standard models.

- Instrument face plates and buttons were changed to all-black, replacing 1990-1991's gray-black. The digital speedometer was relocated above the fuel gauge. Gauge graphics were refined for better legibility.

- The base engine for 1992 was the LT1, a new generation small block. In 1992 Corvettes, the engine developed 300hp (net) at 5000 rpm. Torque was 330 lb.-ft at 4000 rpm. Redline was 5700 rpm, 700 higher than the L98 it replaced. There was an automatic fuel cutoff at 5850 rpm. Power increases were attributed to computer-controlled ignition timing, a low-restriction exhaust system employing two catalytic converters and two oxygen sensors (one converter and one oxygen sensor for each cylinder bank), a higher compression ratio, new camshaft profile, free-flow cylinder heads, and a new multiport fuel injection (MFI) system. At 452 pounds, the 1992 LT1 outweighed the 1991 L98 base engine by twenty-one pounds, due partly to replacement of stainless steel exhaust manifolds with cast iron.

- Corvette's new LT1 engine employed reverse flow cooling, the first such application at Chevrolet. Rather than route coolant from the water pump through the block to the heads, the LT1 routed coolant to the heads first. This permitted higher bore temperatures and reduced ring friction, and helped cooling around the valve seats and spark plug bosses.

- Mobil 1 Synthetic oil was factory fill for the LT1 and Chevrolet recommmended its continued use by owners. An engine oil cooler was no longer available, proven unnecessary when synthetic oil was used.

- Traction control was standard for all 1992 Corvettes. Called Acceleration Slip Regulation (ASR), it was created by Bosch and developed with Corvette engineers. It engaged with the ignition, but could be turned off by an instrument panel switch. ASR used engine spark retard, throttle close down, and brake intervention to limit wheel spin when accelerating. When on and active, a slight accelerator pedal pushback could be felt.

- New Goodyear GS-C tires were introduced as standard equipment on all 1992 Corvettes and were exclusive to Corvettes worldwide for 1992. The GS-C tread design was directional and asymmetrical.

- Improvements in weather sealing were achieved with improved weatherstrip seals. Road noise reduction came from additional insulation in doors and improved insulation over the transmission tunnel.

- The power delay feature was modified so that the passenger door also cut power, in addition to the driver door or fifteen minute time period.

- The 1-millionth Corvette, a white convertible, was built July 2, 1992.

- The Corvette Americana Hall of Fame and Americana Museum, created by Dr. Allen Schery, opened July 31, 1992, in Cooperstown, New York. Citing low tourist interest in the years following the major league baseball strike, and lack of interest from Chevrolet, the museum closed in 1998.

1992 Options

RPO #	DESCRIPTION	QTY	RETAIL $
1YY07	Base Corvette Sport Coupe	14,604	$33,635.00
1YY67	Base Corvette Convertible	5,875	40,145.00
AR9	Base Seats, leather	10,565	475.00
AR9	Base Seats, white leather	752	555.00
AQ9	Sport Seats, leather	7,973	1,100.00
AQ9	Sport Seats, white leather	709	1,180.00
AC1	Power Passenger Seat	16,179	305.00
AC3	Power Driver Seat	19,378	305.00
CC2	Auxiliary Hardtop (convertible)	915	1,995.00
C2L	Dual Removable Roof Panels (coupe)	3,739	950.00
24S	Removable Roof Panel, blue tint (coupe)	6,424	650.00
64S	Removable Roof Panel, bronze tint (coupe)	3,005	650.00
C68	Electronic Air Conditioning Control	18,460	205.00
FX3	Selective Ride and Handling, electronic	5,840	1,695.00
G92	Performance Axle Ratio	2,283	50.00
MN6	6-Speed Manual Transmission	5,487	0.00
NN5	California Emission Requirements	3,092	100.00
UJ6	Low Tire Pressure Warning Indicator	3,416	325.00
UU8	Stereo System, Delco-Bose	3,241	823.00
U1F	Stereo System with CD, Delco-Bose	15,199	1,219.00
V56	Luggage Rack (for convertible)	845	140.00
Z07	Adjustable Suspension Package (coupe)	738	2,045.00
ZR1	Special Performance Package (coupe)	502	31,683.00

• A 350ci, 300hp engine, 4-speed automatic transmission, removable body-color roof panel (coupe) or soft top (convertible), and black cloth seats were included in the base price.

• RPO Z07 included RPO FX3, heavy-duty suspension and heavy-duty brakes. Available with manual or automatic transmission.

• RPO ZR1 included unique bodywork (doors, rear quarters, rear fascia, and rear upper panel) to accept Goodyear Z-rated P315/35ZR17 tires on 11-inch wide rear rims. RPOs AC1, AC3, AQ9, C68, FX3, UJ6, U1F, LT5 (32-valve engine exclusive to the ZR-1), and a specially laminated "solar" windshield were included. (Note: During the model year, RPO AC1 was deleted from the ZR1 package and the price reduced by $305.00.)

• RPO K05 engine block heater was available with base engines and intended for Canada export only. In 1992, 368 were sold at $20.

• Non USA sales in 1992 were 381 Canada, 85 Japan, 500 Europe, 450 Mexico, 24 Gulf States, 10 Saudi Arabia, 1 unregulated country.

1992 Colors

CODE	EXTERIOR	QTY	SOFT TOP	INTERIORS
10	White	4,101	B-Bg-Bk-W	Bk-Lb-Lg-R-W
35	Yellow	678	Bg-Bk-W	Bk-Lb-Lg-W
41	Black	3,209	Bg-Bk-W	Bk-Lb-Lg-R-W
43	Bright Aqua Metallic	1,953	Bg-Bk-W	Bk-Lb-Lg-W
45	Polo Green II Metallic	1,995	Bg-Bk-W	Bk-Lb-W
73	Black Rose Metallic	1,886	Bg-Bk-W	Bk-Lb-Lg-W
75	Dark Red Metallic	1,148	Bg-Bk-W	Bk-Lb-Lg-W
80	Quasar Blue Metallic	1,043	Bg-Bk-W	Bk-Lb-Lg-W
81	Bright Red	4,466	Bg-Bk-W	Bk-Lb-Lg-R-W

• Only interior-exterior combinations shown were considered acceptable.

• Restrictions applied to some soft top and interior color combinations.

• White interiors were not available with coupes early in production.

• Base cloth seats were available only in black. In 1992, 480 were sold.

• Three colors were new for 1992: Bright Aqua Metallic (43), Polo Green II Metallic (45), and Black Rose Metallic (73).

• Convertible top colors sold in 1992 were 2,790 black, 1,892 white, 54 blue, 1,139 beige.

• Interior colors sold in 1992 were 9,186 black, 3,763 light beige, 3,235 red, 2,834 light gray, 1,461 white.

Interior Codes: 103=W/L, 143=Lg/L, 19C=Bk/C, 193=Bk/L, 643=Lb/L, 733=R/L.

Abbreviations: B=Blue, Bg=Beige, Bk=Black, C=Cloth, L=Leather, Lb=Light Beige, Lg=Light Gray, R=Red, W=White.

1993 Corvette

1993 Numbers

Vehicle: 1G1YY23PXP5100001 thru 1G1YY23PXP5121142
1G1YZ23J3P5800001 thru 1G1YZ23J3P5800448 (ZR1)
- For convertibles, sixth digit is a 3.
- Ninth digit is a security code and varies.

Suffix: ZVA: 350ci, 300hp, at   ZVC: 350ci, 405hp
ZVB: 350ci, 300hp, mt

Block: 10125327: 350ci, 300hp   10199001: 350ci, 405hp

Head: 10174389: 350ci, 405hp, lh   10174390: 350ci, 405hp, rh
10205245: 350ci, 300hp

Abbreviations: at=automatic transmission, ci=cubic inch, hp=horsepower, lh=left hand, mt=manual transmission, rh=right hand.

1993 Facts

- Exterior appearance continued virtually unchanged for 1993, but a 40th Anniversary Package (RPO Z25) was optional with all models. The package included a Ruby Red metallic exterior, Ruby Red leather sport seats, power driver seat, special wheel center trim and emblems.

- All leather seats in 1993 Corvettes had "40th" anniversary embroidery in the headrest area. The base black cloth seats did not. This was the last year for the no-cost base cloth seats and 426 1993 Corvettes had them.

- Horsepower for the base LT1 engine remained 300, but three changes made the engine quieter. First, the heat shield design changed from a single-piece stamping to a two-piece sandwich type that was self-damping. Second, new thermoset polyester valve covers with "isolated" mounts replaced 1984-92's magnesium covers. These covers were isolated from the head by a gasket in the normal location, but also by gaskets under the heads of mounting hardware. Third, the LT1 camshaft exhaust lobe profile was modified to reduce the exhaust valve closing velocity. Also, a shortening of the inlet duration permitted more duration for the exhaust so there was no increase in overlap area. Emissions and idle quality weren't adversely affected. A side benefit of closing the inlet valve sooner was an increase in torque from 330 to 340 lb.-ft at 3600 rpm.

- Horsepower increased for the optional ZR1's LT5 engine from 375 to 405hp, a result of modifications to the cylinder heads and valvetrain. Other changes included four-bolt main bearings, a Mobil 1 synthetic oil requirement, platinum-tipped spark plugs, and an electrical, linear exhaust gas recirculation (EGR) system for improved emission control.

- The 1993 Corvette was the first auto sold by GM to feature a passive keyless entry (PKE) system. Working by proximity, a battery-operated key-fob transmitter sent a unique code picked up by a receiver in the Corvette through one of two antennas (in coupes, antennas were in the driver door and rear deck; in convertibles, antennas were in both doors). The transmitter required no specific action by the owner; approaching the vehicle with the transmitter would unlock the doors, turn on the interior light, and disarm the theft-deterrent. Leaving an unlocked vehicle with the transmitter would lock the doors and arm the theft-deterrent. The PKE could be turned off completely and transmitters were programmable for locking and unlocking just the driver door, or both driver and passenger doors. Transmitters for coupes had an extra button for rear hatch release.

- Front wheels for base cars were decreased from 9.5x17 to 8.5x17 and the front tire size from P275/40ZR17 to P255/45ZR17. Rear tire size was increased from P275/40ZR17 to P285/40ZR17. For RPO Z07, 9.5x17 wheels and P275/40ZR17 tires were used front and rear.

- Although the same in design as the previous model, 1993's wheels had a different surface appearance due to a change in finish machining.

- *Car and Driver* magazine's "10 Best" issue (January 1994) determined the 1993 Corvette ZR-1 to be winner of the top speed category at 179mph.

1993 Options

RPO#	DESCRIPTION	QTY	RETAIL $
1YY07	Base Corvette Sport Coupe	15,898	$34,595.00
1YY67	Base Corvette Convertible	5,692	41,195.00
AR9	Base Seats, leather	8,509	475.00
AR9	Base Seats, white leather	766	555.00
AQ9	Sport Seats, leather	11,267	1,100.00
AQ9	Sport Seats, white leather	622	1,180.00
AC1	Power Passenger Seat	18,067	305.00
AC3	Power Driver Seat	20,626	305.00
CC2	Auxiliary Hardtop (convertible)	976	1,995.00
C2L	Dual Removable Roof Panels (coupe)	4,204	950.00
24S	Removable Roof Panel, blue tint (coupe)	6,203	650.00
64S	Removable Roof Panel, bronze tint (coupe)	4,288	650.00
C68	Electronic Air Conditioning Control	19,550	205.00
FX3	Selective Ride and Handling, electronic	5,740	1,695.00
G92	Performance Axle Ratio	2,630	50.00
MN6	6-Speed Manual Transmission	5,330	0.00
NN5	California Emission Requirements	2,401	100.00
UJ6	Low Tire Pressure Warning Indicator	3,353	325.00
UU8	Stereo System, Delco-Bose	2,685	823.00
U1F	Stereo System with CD, Delco-Bose	16,794	1,219.00
V56	Luggage Rack (for convertible)	765	140.00
Z07	Adjustable Suspension Package (coupe)	824	2,045.00
Z25	40th Anniversary Package	6,749	1,455.00
ZR1	Special Performance Package (coupe)	448	31,683.00

• A 350ci, 300hp engine, 4-speed automatic transmission, removable body-color roof panel (coupe) or soft top (convertible), and black cloth seats were included in the base price.

• RPO Z07 included RPO FX3, heavy-duty suspension and heavy-duty brakes. Available with manual or automatic transmission.

• RPO ZR1 included unique bodywork (doors, rear quarters, rear facia, and rear upper panel) to accept Goodyear Z-rated P315/35ZR17 tires on 11-inch wide rear rims. RPOs AC1, AC3, AQ9, C68, FX3, LT5 (32-valve engine exclusive to the ZR-1), U1F, UJ6, and a specially laminated "solar" windshield were included. Available with coupes only.

• RPO K05 engine block heater was available with base engines and intended for Canada export only. In 1993, 406 were sold at $20.

• RPO Z25 40th Anniversary Package included Ruby Red exterior and interior, and special trim. It was available with coupe, convertibles, and RPO ZR1. Domestic sales of Z25 were 6,492 of which 4,204 were standard engine coupes, 2,043 were convertibles, and 245 were ZR-1 coupes. The breakdown of 257 export Z25 models is not known.

• Non USA sales for 1993 included 420 Canada, 121 Japan, 29 Gulf States, 8 unrestricted countries, 193 Europe.

1993 Colors

CODE	EXTERIOR	QTY	SOFT TOP	INTERIORS
10	Arctic White	3,031	Bg-Bk-W	Bk-Lb-Lg-R-W
41	Black	2,684	Bg-Bk-W	Bk-Lb-Lg-R-W
43	Bright Aqua Metallic	1,305	Bg-Bk-W	Bk-Lb-Lg-W
45	Polo Green II Metallic	2,189	Bg-Bk-W	Bk-Lb-W
53	Competition Yellow	517	Bg-Bk-W	Bk-Lb-Lg-W
68	Ruby Red	6,749	Rr	Rr
70	Torch Red	3,172	Bg-Bk-W	Bk-Lb-Lg-R-W
73	Black Rose Metallic	935	Bg-Bk-W	Bk-Lb-Lg-W
75	Dark Red Metallic	325	Bg-Bk-W	Bk-Lb-Lg-W
80	Quasar Blue Metallic	683	Bg-Bk-W	Bk-Lb-Lg-W

• Only interior-exterior combinations shown were considered acceptable.

• Restrictions applied to some soft top and interior color combinations.

• Ruby Red exterior/interior colors were exclusive to the 40th Anniversary Package. Convertibles included a Ruby Red soft top. Other new colors for 1993 were Competition Yellow (53) and Torch Red (70).

• Convertible top color quantities for 1993 were 1,719 black, 996 white, 818 beige, 2,171 ruby red. (Note: total exceeds 1993 production by 12.)

Interior Codes: 103=W/L, 143=Lg/L, 19C=Bk/C, 193=Bk/L, 643=Lb/L, 703=R/L, 793=Rr/L.

Abbreviations: Bg=Beige, Bk=Black, C=Cloth, L=Leather, Lb=Light Beige, Lg=Light Gray, R=Red, Rr=Ruby Red, W=White.

1994 Corvette

Production: 17,984 coupe, 5,346 convertible, 23,330 total

1994 Numbers

Vehicle: 1G1YY22P9R5100001 thru 1G1YY22P9R5122882
1G1YZ22J9R5800001 thru 1G1YZ22J9R5800448 (ZR1)
  • For convertibles, sixth digit is a 3.
  • Ninth digit is a security code and varies

Suffix: ZWA: 350ci, 300hp, at   ZWC: 350ci, 405hp, mt
ZWB: 350ci, 300hp, mt

Block: 10125327: 350ci, 300hp   10199001: 350ci, 405hp

Head: 10174389: 350ci, 405hp, lh, ep   10207643: 350ci, 300hp
10174390: 350ci, 405hp, rh   10225121: 350ci, 405hp, lh

Abbreviations: at=automatic transmission, ci=cubic inch, ep=early production, hp=horsepower, lh=left hand, mt=manual transmission, rh=right hand.

1994 Facts

• After several years of planning and fund raising under the direction of the National Corvette Museum Foundation and its president, the late Dan Gale, the National Corvette Museum opened to the public on September 2, 1994, in Bowling Green, Kentucky.

• The exterior design was carried over from 1993, but two new exterior colors were available, Admiral Blue and Copper Metallic. Copper Metallic availability was limited due to quality concerns and just 116 were sold.

• ZR-1 models received a new, non-directional, five-spoke wheel design. These wheels were unique to ZR-1 and not optional with other models.

• Power output of the base LT1 engine remained 300hp, but several refinements were added. A new sequential fuel injection system improved response, idle quality, driveability and emissions by firing injectors in sequence with the engine's firing order. A more powerful ignition system reduced engine start times, especially in cold temperatures.

• The standard 4-speed automatic transmission was redesigned with electronic controls for improved shift quality and rpm shift-point consistency. Also, a safety interlock was added which required depression of the brake pedal in order to shift from "park."

• A passenger-side airbag and knee bolster, new seat and door trim panel designs, "express down" driver's power window, and a redesigned two-spoke airbag steering wheel were new inside. New white instrument graphics turned to tangerine at night. The tire jack was relocated from the exterior spare tire well to a compartment behind the passenger seat.

• For 1994, all seats were leather as cloth seating was no longer included or optional. Base and optional "sport" styles were available. Both featured less restrictive bolsters to accommodate a wider range of occupant sizes and for improved entry and exit. Controls for base seats with optional power assist were console-mounted with individual controls for driver and passenger. With sport seats, a single set of power assist controls for both seats was console-mounted. Also, individual motors adjusted the lumbar support for sport seats and these controls (and the side bolster control) were relocated from the seat to the console for 1994. Reclining mechanisms for all 1994 seats were manual.

• The rear window for convertibles was changed from plastic to glass and included an in-glass defogger grid.

• RPO FX3 spring rates were lowered to improve ride quality and recommended tire pressures dropped from 35psi to 30psi (except ZR1).

• Air conditioning systems were revised to use R-134a refrigerant, a non-ozone depleting CFC substitute.

• Optional Goodyear Extended Mobility Tires (RPO WY5) had special bead construction to permit use with no air pressure. The low tire pressure warning system (RPO UJ6) was a required option because if the tire was run deflated more than about fifty miles, damage could result. However, the safe driving range was substantially further.

• Twenty-five convertibles were "official cars" for the inaugural NASCAR Brickyard 400 at Indianapolis. Some were later sold to retail customers.

1994 Options

RPO#	DESCRIPTION	QTY	RETAIL $
1YY07	Base Corvette Sport Coupe	17,984	$36,185.00
1YY67	Base Corvette Convertible	5,346	42,960.00
AC1	Power Passenger Seat	17,863	305.00
AC3	Power Driver Seat	21,592	305.00
AQ9	Sport Seats	9,023	625.00
CC2	Auxiliary Hardtop (convertible)	682	1,995.00
C2L	Dual Removable Roof Panels	3,875	950.00
24S	Removable Roof Panel, blue tint (coupe)	7,064	650.00
64S	Removable Roof Panel, bronze tint (coupe)	3,979	650.00
FX3	Selective Ride and Handling, electronic	4,570	1,695.00
G92	Performance Axle Ratio	9,019	50.00
MN6	6-Speed Manual Transmission	6,012	0.00
NG1	New York Emission Requirements	1,363	100.00
UJ6	Low Tire Pressure Warning Indicator	5,097	325.00
U1F	Stereo System with CD, Delco-Bose	17,579	396.00
WY5	Tires, Extended Mobility	2,781	70.00
YF5	California Emission Requirements	2,372	100.00
Z07	Adjustable Suspension Package (coupe)	887	2,045.00
ZR1	Special Performance Package (coupe)	448	31,258.00

• A 350ci, 300hp engine, 4-speed automatic transmission, removable body-color roof panel (coupe) or soft top (convertible), Delco stereo system with cassette, and leather seats were included in the base price.

• Preferred Equipment Group One included electronic air conditioning control, Delco-Bose stereo system with cassette, and power driver seat (RPO AC3) for $1,333.00.

• RPO G92 was available without restriction.

• RPO WY5 required RPO UJ6; not available with RPOs Z07 or ZR1.

• RPO Z07 included RPO FX3, heavy-duty suspension and heavy-duty brakes. Available with manual or automatic transmission.

• RPO ZR1 included unique bodywork (doors, rear quarters, rear fascia, and rear upper panel) to accept Goodyear Z-rated P315/35ZR17 tires on 11-inch wide rear wheels (front and rear wheel design for 1994 was new and unique to the ZR-1). Electronic air conditioning control and RPOs AC1, AC3, AQ9, C68, FX3, LT5 (32-valve engine exclusive to the ZR-1), U1F, UJ6, and a specially laminated "solar" windshield were included. Available with coupes only; availability limited.

• RPO K05 engine block heater was available with base engines and intended for Canada export only. Sales for 1994 were 493 at $20.

• Non USA sales for 1994 were 513 Canada, 79 Japan, 50 Germany, 31 Switzerland, 16 Austria, 8 Belgium, 6 Gulf States, 2 Luxembourg, 1 Netherlands, 1 France.

1994 Colors

CODE	EXTERIOR	QTY	SOFT TOP	INTERIORS
10	Arctic White	4,066	Bg-Bk-W	Bk-Lb-Lg-R
28	Admiral Blue	1,584	Bg-Bk-W	Bk-Lb-Lg
41	Black	4,136	Bg-Bk-W	Bk-Lb-Lg-R
43	Bright Aqua Metallic	1,209	Bg-Bk-W	Bk-Lb-Lg
45	Polo Green Metallic	3,534	Bg-Bk	Bk-Lb
53	Competition Yellow	834	Bg-Bk-W	Bk-Lb-Lg
66	Copper Metallic	116	Bg-Bk-W	Bk-Lb-Lg
70	Torch Red	5,073	Bg-Bk-W	Bk-Lb-Lg-R
73	Black Rose Metallic	1,267	Bg-Bk-W	Bk-Lb-Lg
75	Dark Red Metallic	1,511	Bg-Bk-W	Bk-Lb-Lg

• Exterior, interior and soft top combinations were recommended as most desirable, but other combinations could be ordered.

• Admiral Blue (28) and Copper Metallic (66) were new colors for 1994. All others carried over from 1993.

• All interiors were leather and all interior colors were available in both base and sport seats. Cloth was not available.

• Convertible top color quantities for 1994 were 2,770 black, 1,079 white, 1,471 beige. (Note: production exceeds total by 26).

Interior Codes: 143=Lg/L, 193=Bk/L, 643=Lb/L, 703=R/L.

Abbreviations: Bg=Beige, Bk=Black, Lb=Light Beige, Lg=Light Grey, R=Red, W=White.

1995 Corvette

1995 Numbers

Vehicle: 1G1YY22P7S5100001 thru 1G1YY22P7S5120294
1G1YZ22J0S5800001 thru 1G1YZ22J0S5800448 (ZR1)
- For convertibles, sixth digit is a 3.
- Ninth digit is a security code and varies.

Suffix: ZUC: 350ci, 300hp, at   ZUF: 350ci, 300hp, mt
ZUD: 350ci, 405hp, mt

Block: 10125327: 350ci, 300hp   10199001: 350ci, 405hp

Head: 10174390: 350ci, 405hp,rh, ep   10225121: 350ci, 405hp, lh
10207643: 350ci, 300hp   10225122: 350ci, 405hp, rh

Abbreviations: at=automatic transmission, ci=cubic inch, ep=early production, hp=horsepower, lh=left hand, mt=manual transmission, rh=right hand.

1995 Facts

- The 1995 exterior was distinguished from 1994 by restyling of the front fender "gill" air vents. A new exterior color, Dark Purple Metallic, was added, but 1994's Copper Metallic and Black Rose Metallic were deleted.

- Corvette paced the Indianapolis 500 race in 1995 and produced for sale a Dark Purple and White convertible pace car model. On January 3, 1995, Chevrolet announced to its dealers that 522 pace cars would be produced with 87 of these for Indianapolis festival and public relations use and 20 for Canada and export. The remaining 415 would go one each to the top 415 dealers. Actual final production was 527.

- Optional Sport Seats had stronger "French" seam stitching. A readout for automatic transmission fluid temperature was added to the instrument display. Out of sight were numerous Velcro straps to reduce rattles, and a stronger radio mount for less CD skipping. A drip tube was designed into the A-pillar weatherstrip for improved water intrusion control.

- The base LT1 engine continued with the same 300hp and 340 lb-ft torque ratings, but there were refinements. Late in 1994 production, connecting rods were changed to a powdered-metal design to improve both strength and weight uniformity. Fuel injectors were revised to better cope with alcohol-blend fuels and to reduce fuel dripping after engine shutdown. The engine cooling fan was modified for quieter operation.

- This was the ZR-1's last year. Mercury Marine in Stillwater, Oklahoma, completed all LT5 engines in November 1993. Tooling, owned by GM, was removed from Mercury Marine's factory and all engines, specially sealed, were shipped to Corvette's Bowling Green assembly plant for storage until needed. Before September 1, 1993, all internal engine warranty repair was done by Mercury Marine. Between September 1, 1993 and December 31, 1993, internal repairs were done by Mercury Marine if engines had under 12,000 miles or 12 months service. Chevrolet handled service not performed by Mercury Marine, including all after January 1, 1994. Total 1995 ZR-1 production was predetermined at 448 units, the same as 1993 and 1994. Total ZR-1 production for 1990 through 1995 was 6,939.

- Clutch controls in the four-speed automatic transmission were improved for smoother shifting, and its torque converter was both lighter and stronger. The 6-speed manual was redesigned by replacement of the reverse lockout with a high-detent design for easier operation.

- The larger brake package, included previously with Z07 and ZR-1 performance options, was included for 1995 with all models. And all 1995's had the latest anti-lock/traction control (ABS/ASR-5) system.

- The extended mobility "run flat" tires introduced as a 1994 option minimized the need for a spare tire. So 1995's RPO N84 created a delete spare option which reduced weight and included a credit of $100.00.

- Base suspension models had lower front and rear spring rates.

- Windshield wiper arms had revised contact angles and higher contact force to reduce chatter at all speeds, and lift at high speeds.

1995 Options

RPO#	DESCRIPTION	QTY	RETAIL $
1YY07	Base Corvette Sport Coupe	15,771	$36,785.00
1YY67	Base Corvette Convertible	4,971	43,665.00
AG1	Power Driver Seat	19,012	305.00
AG2	Power Passenger Seat	15,323	305.00
AQ9	Sport Seats	7,908	625.00
CC2	Auxiliary Hardtop (convertible)	459	1,995.00
C2L	Dual Removable Roof Panels (coupe)	2,979	950.00
24S	Removable Roof Panel, blue tint (coupe)	4,688	650.00
64S	Removable Roof Panel, bronze tint (coupe)	2,871	650.00
FX3	Selective Ride and Handling, electronic	3,421	1,695.00
G92	Performance Axle Ratio	10,056	50.00
MN6	6-Speed Manual Transmission	4,784	0.00
NG1	New York Emission Requirements	268	100.00
N84	Spare Tire Delete	418	-100.00
UJ6	Low Tire Pressure Warning Indicator	5,300	325.00
U1F	Stereo System with CD, Delco-Bose	15,528	396.00
WY5	Tires, Extended Mobility	3,783	70.00
YF5	California Emission Requirements	2,026	100.00
Z07	Adjustable Suspension Package (coupe)	753	2,045.00
Z4Z	Indy 500 Pace Car Replica (convertible)	527	2,816.00
ZR1	Special Performance Package (coupe)	448	31,258.00

• A 350ci, 300hp engine, 4-speed automatic transmission, removable body-color roof panel (coupe) or soft top (convertible), Delco stereo system with cassette, and leather seats were included in the base price.

• Preferred Equipment Group One included electronic air conditioning control, Delco-Bose stereo system with cassette, and power driver seat (RPO AG1) for $1,333.00.

• RPO WY5 required RPO UJ6; not available with RPOs Z07 or ZR1.

• RPO Z07 included RPO FX3, special springs, stabilizers and bushings, 17x9 1/2" wheels and P275/40ZR17/N black-letter tires. Required AG1 and AG2. Available with manual or automatic transmission, but required G92 with automatic transmission..

• RPO Z4Z (convertible only) included special Dark Purple and Arctic White paint, white convertible top, special graphics and trim.

• RPO ZR1 included unique bodywork (doors, rear quarters, rear fascia, and rear upper panel) to accept Goodyear Z-rated P315/35ZR17 tires on special-design 11-inch wide rear wheels. Electronic air conditioning control and RPOs AG1, AG2, AQ9, C68, FX3, LT5 (32-valve engine exclusive to the ZR-1), U1F, UJ6, and a specially laminated "solar" windshield were also standard. Available with coupes only.

• Non USA sales for 1995 were 554 Canada, 11 Mexico, 251 export.

1995 Colors

CODE	EXTERIOR	QTY	SOFT TOP	INTERIORS
05	Dark Purple Metallic	1,049	Bg-Bk-W	Bk-Lb-Lg
05/10	Dark Purple/White	527	W	BkPc
10	Arctic White	3,381	Bg-Bk-W	Bk-Lb-Lg-R
28	Admiral Blue	1,006	Bg-Bk-W	Bk-Lb-Lg
41	Black	3,959	Bg-Bk-W	Bk-Lb-Lg-R
43	Bright Aqua Metallic	909	Bg-Bk-W	Bk-Lb-Lg
45	Polo Green Metallic	2,940	Bg-Bk	Bk-Lb
53	Competition Yellow	1,003	Bg-Bk-W	Bk-Lb-Lg
70	Torch Red	4,531	Bg-Bk-W	Bk-Lb-Lg-R
75	Dark Red Metallic	1,437	Bg-Bk-W	Bk-Lb-Lg

• Other combinations could be ordered, except for Pace Car convertible replica which could not deviate from combinations shown.

• All interiors were leather and all interior colors were available in both base and sport seats.

• ZR-1 color quantities were 41 Arctic White, 12 Admiral Blue, 121 Black, 13 Bright Aqua, 31 Polo Green, 49 Competition Yellow, 140 Torch Red, 25 Dark Purple, 16 Dark Red.

• Convertible top color quantities for 1995 were 2,505 black, 1,222 white, 1,244 beige.

Interior Codes: 143=Lg, 193=Bk, 194=Bk-Pc, 643=Lb, 703=R.

Abbreviations: Bg=Beige, Bk=Black, Lb=Light Beige, Lg=Light Gray, Pc=Pace Car Replica, R=Red, W=White.

1996 Corvette

Production: 17,167 coupe, 4,369 convertible, 21,536 total

1996 Numbers

Vehicle: 1G1YY2257T5100001 thru 1G1YY2251T5120536
1G1YY2251T5600001 thru 1G1YY2251T5601000 (Grand Sport)
• For convertibles, sixth digit is a 3.
• Eighth digit is a P for LT1, 5 for LT4.
• Ninth digit is a security code and varies.

Suffix: ZXA: 350ci, 300hp, at (LT1)   ZXD: 350ci, 330hp, mt (LT4)

Block: 10125327: all

Head: 10207643: 350ci, 300hp, at   12551561: 350ci, 300hp, at, lp
10239902: 350ci, 330hp, mt, ep   12555690: 350ci, 330hp, mt

Abbreviations: at=automatic transmission, ci=cubic inch, ep=early production, hp=horsepower, lp=late production, mt=manual transmission.

1996 Facts

• A new version of Chevy's 350-cubic-inch small block, RPO LT4, became optional exclusively with 1996 Corvettes. Rated at 330-horsepower, 30 more than the base LT1, the LT4 had higher compression (10.8:1 vs 10.4:1), new aluminum head design, Crane roller rocker arms, revised camshaft profile, and other major and minor tweaks. The LT4's redline increased to 6300 rpm (5700 rpm for LT1), so LT4-equipped models had 8000 rpm tachometers instead of the base 6000 rpm. LT4 was available with all Corvette models, but only with manual transmissions.

• LT1 engines were mated only to automatic transmissions which had improved friction materials for the intermediate clutch and front/rear bands, improved shift quality and more durable torque converters.

• RPO Z16 Grand Sport included LT4 engine, Admiral Blue paint with white center stripe, and special detailing. The previous year's ZR-1-style five-spoke 17" wheels were used, but painted black. Like the ZR-1, tires for Grand Sport coupes were P275/40ZR17 front and P315/35ZR17 rear. But Grand Sport coupes had added rear fender flares rather than ZR-1's wider rear panels. Convertible Grand Sport tires were P255/45ZR17 front, P285/40ZR17 rear, with no rear fender flares. Interior choices were limited to black, or a red-black combination. Corvettes with the Grand Sport option had separate serial number sequences.

• LT1 and LT4 engines had a new throttle body for 1996. Those for LT4 engines had red "Grand Sport" lettering, regardless of the application.

• "Collector Edition" (RPO Z15) included Sebring Silver paint and special trim. ZR-1-style 17" five-spoke wheels were used, but painted silver with P255/45ZR17 front and P285/40ZR17 rear tires. Black, red or gray interiors were available, but soft top color was limited to black.

• RPO F45, Selective Real Time Damping, was priced the same as 1995's FX3 Selective Ride option ($1,695), but was substantially different. Using data from wheel travel sensors and the Powertrain Control Module, a controller calculated the damping mode that would provide optimum control via special shock absorbers. It could alter each shock individually (unlike the earlier system which changed all shocks simultaneously) every 10 to 15 milliseconds, or about every foot of roadway at 60 mph.

• Performance Handling Package (RPO Z51) previously optional from 1984 thru 1990, returned in 1996 with different content but similar intent. It included Bilstein shock absorbers with stiffer springs and thicker stabilizer bars. If ordered with an automatic transmission, a 3.07:1 axle was required. Tires were P275/40ZR17 on 17x9.5" aluminum wheels, except for Z51 Grand Sports which had P315/35ZR17 rear tires on 17x11" wheels. Z51 was limited to coupes.

• 1996's On-Board-Diagnostics were much more sophisticated and complex, the number of diagnostic codes increasing from 60 to 140.

• Even though it was the last of a generation, the 1996 Corvette regained the top spot (after two years in second place) in *AutoWeek* magazine's annual subscriber survey of American cars in which readers had the most pride. Dodge Viper was second. (*AutoWeek* July 1, 1996).

1996 Options

RPO#	DESCRIPTION	QTY	RETAIL $
1YY07	Base Corvette Sport Coupe	17,167	37,225.00
1YY67	Base Corvette Convertible	4,369	45,060.00
AG1	Power Driver Seat	19,798	305.00
AG2	Power Passenger Seat	17,060	305.00
AQ9	Sport Seats	12,016	625.00
CC2	Auxiliary Hardtop (convertible)	429	1,995.00
C2L	Dual Removable Roof Panels (coupe)	3,983	950.00
24S	Removable Roof Panel, blue tint (coupe)	6,626	650.00
64S	Removable Roof Panel, bronze tint (coupe)	2,492	650.00
F45	Selective Real Time Damping, electronic	2,896	1,695.00
G92	Performance Axle Ratio	9,801	50.00
LT4	350 cubic-inch, 330 horsepower Engine	6,359	1,450.00
MN6	6-Speed Manual Transmission	6,359	0.00
N84	Spare Tire Delete	986	-100.00
UJ6	Low Tire Pressure Warning Indicator	6,865	325.00
U1F	Compact Disc Delco-Bose (reqs PEG 1)	17,037	396.00
WY5	Tires, Extended Mobility	4,945	70.00
Z15	Collector Edition	5,412	1,250.00
Z16	Grand Sport Package ($2,880 w/convertible)	1,000	3,250.00
Z51	Performance Handling Package	1,869	350.00

• A 350ci, 300hp engine, 4-speed automatic transmission, removable body-color roof panel (coupe) or soft top (convertible), Delco stereo system with cassette, and leather seats were included in base prices.

• RPO LT4 included (and was available only with) RPO MN6.

• Preferred Equipment Group (PEG) 1 included electronic air conditioning control, Delco-Bose stereo with cassette, and RPO AG1 for $1,333.00.

• RPO F45 required RPOs AG1 and AG2.

• RPO WY5 required RPO UJ6; not avail with RPO Z16 coupe or Z51.

• RPO Z15 included 17" 5-spoke wheels painted silver, black brake calipers with silver "Corvette" lettering, Sebring Silver exterior paint, special emblems, P255/45ZR17 front tires, P285/40ZR17 rear tires, and perforated sports seats with "Collector Edition" embroidery. Total production of 5,412 was divided 4,031 coupe, 1,381 convertible.

• RPO Z16 included Admiral Blue paint with white stripe, red left-fender hash marks, 17" 5-spoke aluminum wheels painted black, black brake calipers with silver "Corvette" lettering, P275/40ZR/17 front and P315/35ZR/17 rear tires (coupe), rear wheel flares (coupe), black floor mats, and perforated sport seats with "Grand Sport" embroidery. Total production of 1,000 was divided 810 coupe, 190 convertible.

• RPO Z51 had Bilstein shocks, special front and rear springs, stabilizers and bushings, 17x9.5"wheels with P275/40 ZR 17 tires. (Z16 with Z51 had 17x11" rear wheels with P315/35ZR17 tires.) Z51 required AG1 and AG2. Available with manual or automatic, but automatic required G92.

• Non USA sales for 1996 were 348 Canada, 11 Mexico, 172 export.

1996 Colors

CODE	EXTERIOR	QTY	SOFT TOP	INTERIORS
05	Dark Purple Metallic	320	Bg-Bk-W	Bk-Lb-Lg
10	Arctic White	3,210	Bg-Bk-W	Bk-Lb-Lg-R
13	Sebring Silver Metallic	5,412	Bk	Bk-Lg-R
28	Admiral Blue	1,000	W	Bk-R/Bk
41	Black	3,917	Bg-Bk-W	Bk-Lb-Lg-R
43	Bright Aqua Metallic	357	Bg-Bk-W	Bk-Lb-Lg
45	Polo Green Metallic	2,414	Bg-Bk	Bk-Lb
53	Competition Yellow	488	Bg-Bk-W	Bk-Lb-Lg
70	Torch Red	4,418	Bg-Bk-W	Bk-Lb-Lg-R

•Other combinations could be ordered, except for codes 13 and 28 which could not deviate from combinations shown.

•All seating was leather and all interior colors were available in base and sport seats except for codes 13 and 28 which were sport seat only.

• Grand Sport interior quantities were 730 black, 270 red/black.

Interior Codes: 143=Lg, 144=Lg-Ce, 193=Bk, 194=Bk-Ce, 195=Bk-Gs, 643=Lb, 703=R, 704=R-Ce, 705=R/Bk-Gs.

Abbreviations: Bg=Beige, Bk=Black, Ce=Collector Edition, Gs=Grand Sport, Lb=Light Beige, Lg=Light Gray, R=Red, R/Bk=Red and Black, W=White.

1997 Corvette

Production: 9,752 coupes

1997 Numbers

Vehicle: 1G1YY22G1V5100001 thru 1G1YY22G9V5109707
- Eighth digit is engine code: G=346ci, 345hp (LS1)
- Ninth digit is a security code and varies.
- Ending VIN did not match production quantity because production counted all vehicles built during 1997 including pilots and prototypes.

Suffix: ZYC: 346ci, 345hp, mt   ZYD: 346ci, 345hp, at

Block: 12550592: 346ci, 345hp

Head: 10215339: 346ci, 345hp   12558806: 346ci, 345 hp, lp

Abbreviations: at=automatic transmission, ci=cubic inch, hp=horsepower, lp=late production, mt=manual transmission.

1997 Facts

- Unlike past "all-new" models which had carryover engines and drivelines, the 1997 Corvette had a completely new engine driving a rear transaxle, a Corvette first. Virtually all interior, exterior and suspension components were redesigned for this vehicle.

- The rear transaxle combined a GM-built, electronically-controlled four-speed automatic transmission, or a Borg-Warner six-speed manual with Computer-Aided Gear Selection (CAGS) with a limited-slip axle built by Getrag. Locating the transmission in the rear maintained nearly equal front-to-rear weight distribution (51.4/48.6 with automatic), and improved interior space by enlarging the footwells.

- Wheelbase increased from 96.2" to 104.5", length from 178.5" to 179.7", width from 70.7" to 73.6", height from 46.3" to 47.8," track by 4.4" in front and 2.9" at the rear. Moving the wheels more to the corners increased stability and interior space. Weight was reduced by about eighty pounds.

- The 1997 interior featured large analog speedometer and tachometer gauges flanked by secondary instruments, with "black" lighting. Seats, by Lear Corporation, were a new design, available in standard or optional sport. The handbrake was relocated from left of the driver to the center console. Stepover height was reduced by over three inches.

- The LS1 was built by GM's engine plant in Romulus, Michigan. Most previous Corvette small-block V8's were built in Flint, Michigan.

- LS1 had the same 4.4" bore spacing and similar displacement (346ci) as the 350ci engine family it replaced, but otherwise was new and state-of-the-art in pushrod V8 design. The block was closed-deck aluminum alloy with cast-in cylinder liners. The block had a deep skirt that extended past the main bearing caps. Four-bolt mains were also cross-bolted to the block for rigidity. The shallow cast-aluminum oil pan had side reservoirs, and was a structural member of the engine. Output was 345hp at 5600 rpm, torque 350 lb-ft. at 4400 rpm and redline at 6000 rpm. Premium fuel was recommended and LS1's came filled with 5w30 Mobil 1 synthetic oil. An assembled LS1 was 44 lbs lighter than 1996's LT4.

- LS1's composite intake manifold saved weight and improved air flow. Aluminum valve covers had separate ignition coils mounted close to each spark plug. Spark timing signals were sent by crankshaft and camshaft sensors. Dual knock sensors were mounted under the intake manifold. Firing order was revised from 1-8-4-3-6-5-7-2 to 1-8-7-2-6-5-4-3.

- Engine heads were cast alloy with identical ports. The camshaft was hollow for reduced mass. The exhaust manifold was two-layer stainless steel with an air space insulator to more quickly warm the catalytic converter for reduced start-up emissions.

- Wheel size was 17x8.5" front, 18x9.5" rear. Tire size was P245/45ZR17 front, P275/40ZR18 rear. Tires were extended mobility (run flat) with no provision for a spare. Suspension had short-long arm (SLA) design at each corner, also known as double wishbone or double A-arm. Springs were composite plastic front and rear, transversely mounted.

1997 Options

RPO#	DESCRIPTION	QTY	RETAIL $
1YY07	Base Corvette Sport Coupe	9,752	$37,495.00
AAB	Memory Package	6,186	150.00
AG2	Power Passenger Seat	8,951	305.00
AQ9	Sport Seats	6,711	625.00
B34	Floor Mats	9,371	25.00
B84	Body Side Moldings	4,366	75.00
CC3	Removable Roof Panel, blue tint (coupe)	7,213	650.00
C2L	Dual Removable Roof Panels	416	950.00
CJ2	Electronic Dual Zone Air Conditioning	7,999	365.00
D42	Luggage Shade and Parcel Net	8,315	50.00
F45	Selective Real Time Damping, electronic	3,094	1,695.00
G92	Performance Axle Ratio (Automatic only)	2,739	100.00
MN6	6-Speed Manual Transmission	2,809	815.00
NG1	Massachusetts/New York Emissions	677	170.00
T96	Fog Lamps	8,829	69.00
UN0	Delco Stereo System with CD	6,282	100.00
U1S	Remote Compact 12-Disc Changer	4,496	600.00
V49	Front License Plate Frame	2,258	15.00
YF5	California Emissions	885	170.00
Z51	Performance Handling Package	1,077	350.00

• A 346ci, 5.7L 345hp engine (LS1), 4-speed automatic transmission, removable body-color roof panel, Delco stereo cassette with Bose speakers, and leather seats were all standard features.

• RPO AAB (Memory Package) remembered outside-rear-view-mirror, radio, heater-vent-air conditioning control, and power driver seat settings. RPO CJ2 was required.

• RPO AQ9 (Sport Seats) required RPO AG2 (Power Passenger Seat).

• RPO B84 had two pieces per side, and thickness tapered front to rear.

• RPO C2L (Dual Removable Roof Panels) included standard body-color roof panel plus blue-tint transparent panel.

• RPO D42 was a thin fabric cover which attached to the hatch. If not desired, it could be kept in a storage bag which was included.

• RPO F45 (Selective Real Time Damping) included driver-adjustable ride control system. Not available with RPO Z51.

• RPO G92 (Performance Axle Ratio) provided 3.15:1 ratio with automatic transmission (2.73:1 standard with automatic transmission; 3.42:1 standard with manual transmission).

• RPO U1S (Remote CD Changer) could be ordered with standard Delco cassette stereo system or with RPO UN0 (Delco Stereo System with CD).

• RPO V49 was a U-shaped plastic trim piece which the license slipped into. It was an appearance item, not required for front license mounting.

• RPO Z51 (Performance Handling Package) included stiffer springs and stabilizer bars. Required RPO G92 with automatic transmission; not available with RPO F45.

• Non-US sales included 46 to Mexico, 325 to Canada, and 426 export.

• Nine (9) buyers took delivery at the Bowling Green plant (RPO R8C).

1997 Colors

CODE	EXTERIOR	QTY	INTERIORS
10	Arctic White	1,341	B-Lg-R
13	Sebring Silver Metallic	2,164	B-Lg-R
23	Nassau Blue Metallic	292	B-Lg-R
41	Black	2,393	B-Lg-R
53	Light Carmine Red Metallic	381	B-Lg
70	Torch Red	3,026	B-Lg-R
87	Fairway Green Metallic	155	B-Lg-R

• Exterior-interior combinations were recommended as most desirable, but any combination could be ordered.

• Exterior codes 10, 13, 41, 70 and 87 were the only colors available early.

• All seating was leather and all interior colors were available in both base and Sport Seats.

• Interiors sold in 1997 were 6,481 Black, 2,543 Light Gray, 728 Red.

Interior Codes: 193=B, 923=Lg, 943=R.

Abbreviations: B-Black, Lg-Light Gray, R=Firethorn Red.

1998 Corvette

Production: 19,235 coupe, 11,849 convertible, 31,084 total

1998 Numbers

Vehicle: 1G1YY32G2W5100001 thru 1G1YY22G2W5131069
- Sixth digit is a 2 for coupe, 3 for convertible.
- Eighth digit is engine code: G=346ci, 345hp (LS1)
- Ninth digit is a security code and varies.
- Ending VIN did not match production quantity because production counted all vehicles built during 1998 including pilots and prototypes.

Suffix: ZZC: 346ci, 345hp, at   ZZD: 346ci, 345hp, mt

Block: 12550592: 346ci, 345hp   12559846: 346ci, 345hp, rlp

Head: 12558806: 346ci, 345hp

Abbreviations: at=automatic transmission, ci=cubic inch, hp=horsepower, mt=manual transmission, rlp=random late production.

1998 Facts

• After a one-model-year absence, a convertible was again available in 1998. The 1998 convertible weighed 114 pounds less than 1996's, yet was more than four times stiffer torsionally.

• Lowering the convertible's soft top required release of two front latches, then pushing a button for release of a hard tonneau. If side windows were up, pushing the button also lowered them. The tonneau pivoted up and back, and the top folded manually into its storage bay. Raising the top was a reversal of these steps. Articulated geometry of the bows eliminated rear latches. The rear window was glass and heated.

• For the first time since 1962, a separate trunk with outside access returned in 1998. It was a standard feature of the convertible body style; coupes retained their hatchback design as before. The convertible's storage space was just slightly more than half that of coupes (13.9 cubic feet compared to 24.8), but it was larger than virtually all competing soft top two-seaters, and twice that of the previous Corvette generation's convertible. Even with the top folded down, storage space was 11.1 cubic feet, adequate for two golf bags.

• Active Handling System, RPO JL4, was introduced interim and provided additional control in slippery conditions or in extreme handling situations. It countered oversteer or understeer by selectively applying individual wheel brakes. Throttle was not affected.

• Coefficient of drag was .33 for convertibles with top up (.29 for coupes).

• An Indianapolis 500 Pace Car replica convertible included Pace Car Purple exterior with black and yellow leather interior, black convertible top, yellow-painted stock wheels, and special accents. These were fully equipped. The only available options were front license plate frame, a 12-disc CD changer, body-side moldings and magnesium wheels.

• Noise reduction measures included a revised accessory drive tensioner, removal of the alternator rear brace to eliminate high-rpm whine, and clips to better hold window glass to seals during high-speed operation. A quieter electric fuel pump was used after the start of 1998 production.

• Steering castor angle was increased for improved tracking and Selective Real Time Damping (RPO F45) was modified for better wheel control.

• The transmission cooler changed from copper-nickel to stainless steel.

• The multi-piece cooling fan shroud was replaced by a one-piece unit.

• A second-gear start select mode was added to automatic transmissions to limit wheel spin on slippery surfaces.

• For non-Pace Car models, all exterior colors carried over, and two new exterior colors were added for 1998, Light Pewter Metallic and Medium Purple Pearl Metallic. Light Oak became the fourth interior color choice, added to 1997's Black, Light Gray, and Red. Convertible top color choices were Black or Light Oak cloth, and White vinyl, all with black liners.

• Automotive media praised the Corvette, voting it *North American Car of the Year* in January at the 1998 North American International Auto Show in Detroit. *AutoWeek*, January 12, 1998.

1998 Options

RPO#	DESCRIPTION	QTY	RETAIL $
1YY07	Base Corvette Sport Coupe	19,235	$37,495.00
1YY67	Base Corvette Convertible	11,849	44,425.00
AAB	Memory Package	24,234	150.00
AG2	Power Passenger Seat	28,575	305.00
AQ9	Sport Seats	22,675	625.00
B34	Floor Mats	30,592	25.00
B84	Body Side Moldings	17,070	75.00
C2L	Dual Removable Roof Panels (coupe)	5,640	950.00
CC3	Removable Roof Panel, blue tint (coupe)	6,957	650.00
CJ2	Dual Zone Air Conditioning	26,572	365.00
D42	Luggage Shade and Parcel Net (coupe)	16,549	50.00
F45	Selective Real Time Damping, electronic	8,374	1,695.00
G92	Performance Axle (3.15 ratio for automatic)	13,331	100.00
JL4	Active Handling System	5,356	500.00
MN6	6-Speed Manual Transmission	7,106	815.00
NG1	Massachusetts/New York Emissions	2,701	170.00
N73	Magnesium Wheels	1,425	3,000.00
T96	Fog Lamps	29,310	69.00
UN0	Delco Stereo System with CD	18,213	100.00
U1S	Remote Compact 12-Disc Changer	16,513	600.00
V49	Front License Plate Frame	16,087	15.00
YF5	California Emissions	3,111	170.00
Z4Z	Indy Pace Car Replica ($5,804 w/manual)	1,163	5,039.00
Z51	Performance Handling Package	4,249	350.00

• A 346ci, 345hp engine, 4-speed automatic transmission, removable body-color roof panel (coupe) or soft top (convertible), Delco stereo cassette with Bose speakers, and leather seats were standard features.

• RPO AAB (Memory Package) remembered outside-rear-view-mirror, radio, hvac, and power driver seat settings. RPO CJ2 was required.

• RPO AQ9 (Sport Seats) required RPO AG2 (Power Passenger Seat).

• RPO C2L (Dual Removable Roof Panels) included standard body-color roof panel plus blue-tint transparent panel.

• RPO F45 (Selective Real Time Damping) included driver-adjustable ride control system. Not available with RPO Z51.

• RPO G92 quantity was the number with automatic transmission. Overall axle ratio distribution was 10,562 2.73:1, 13,415 3.15:1, 7,106 3.42:1.

• RPO N73 limited mainly to export. Domestic sales were 233 sets.

• RPO U1S (Remote CD Changer) could be ordered with standard Delco cassette stereo system or with RPO UN0.

• RPO Z4Z (convertible) included Pace Car Purple exterior with black/yellow leather interior, special trim; automatic or manual transmission. Total sales of 1,163 were split 616 automatic transmission, 547 manual.

• RPO Z51 (Performance Handling Package) included stiffer springs and stabilizers. Required RPO G92 with automatic; not available with F45.

1998 Colors

CODE	EXTERIOR	QTY	SOFT TOP	INTERIORS
10	Arctic White	3,346	B-Lo-W	B-Lg-Lo-R
11	Light Pewter Metallic	3,276	B-W	B-Lg-R
13	Sebring Silver Metallic	4,637	B-W	B-Lg-R
21	Pace Car Purple	1,163	B	Y
23	Nassau Blue Metallic	1,098	B-Lo-W	B-Lg-Lo
28	Navy Blue Metallic	14	B-Lo-W	B-Lg-Lo
41	Black	6,597	B-Lo-W	B-Lg-Lo-R
53	Light Carmine Red Metallic	1,567	B-Lo-W	B-Lg-Lo-R
58	Aztec Gold	15	B-Lo-W	B-Lg-Lo
70	Torch Red	8,767	B-Lo-W	B-Lg-Lo
87	Fairway Green Metallic	223	B-Lo-W	B-Lg-Lo
95	Medium Purple Pearl Metallic	381	B-Lo-W	B-Lg-Lo

• Other combos could be ordered, except #21 which couldn't deviate.

• Interior colors available in base or Sport, except Yellow (Sport only).

• Codes 28 and 58 were sold only as used cars by Chevrolet.

• Convertible tops for 1998 were 8,630 black, 843 white, 2,376 light oak.

Interior Codes: 193=B, 194=Y, 673=Lo, 923=Lg, 943=R.

Abbreviations: B-Black, Lg=Light Gray, Lo=Light Oak, R=Red, Y=Yellow.

1999 Corvette

18,078 coupe, 11,161 convertible, 4,031 hardtop, 33,270 total

1999 Numbers

Vehicle: 1G1YY12G5X5100001 thru 1G1YY22G0X5133283
- Sixth digit is a 1 for hardtop, 2 for coupe, 3 for convertible.
- Eighth digit is engine code: G=346ci, 345hp (LS1)
- Ninth digit is a security code and varies.
- Ending VIN did not match production quantity because production counted pilots and prototypes.

Suffix: ZAB: 346ci, 345hp, at   ZAA: 346ci, 345hp, mt

Block: 12559846 & 12550592: 346ci, 345hp, ep
12560621 & 12562174: 346ci, 345hp
12560626 & 12559378: 346ci, 345hp, lp

Head: 12559853: 346ci, 345hp

Abbreviations: at=automatic transmission, ci=cubic inch, ep=early production, hp=horsepower, lp=late production, mt=manual transmission.

1999 Facts

• A hardtop model joined the lineup for 1999, the last of a three-model strategy planned for the fifth generation. The fixed-roof model followed the 1997 targa-top hatchback (coupe), and the 1998 convertible, and was to have less standard equipment and lower pricing. It was Corvette's lowest-cost model, but earlier plans for cost-saving features such as cloth seats, manual locks, and smaller wheels and tires were cancelled.

• The hardtop included the 345-hp LS1 engine, but was limited to manual transmission which was included in its price. Base seats in black leather were standard. Sport seats and other interior colors weren't available. Exterior color choices were limited to White, Light Pewter, Nassau Blue, Black, and Torch Red. Z51 suspension was included. Active Handling (JL4) was optional, but Selective Real Time Damping (F45) wasn't.

• The hardtop model was not a version of the coupe less targa. It was more similar to a convertible with a hardtop permanently bolted and bonded in place. It had the same external trunk as the convertible. The hardtop model had the highest chassis stiffness of all three Corvette models, some 12% stiffer than a coupe with targa in place. The hardtop was Corvette's lightest model by about eighty pounds.

• A Power Telescoping Steering Column (RPO N37) became optional in 1999 for coupes and convertibles, but not hardtop models. It offered plus or minus 20mm of travel from the fixed wheel position. The tilt feature retained manual control and was standard with all models.

• Steering hardware was modified for extra sensitivity and less "wander."

• A new sill trim plate was added to all 1999 models.

• Twilight Sentinel (RPO T82) was another new option for 1999, but limited to coupe and convertible models. The system used a low-light sensor to automatically lift the headlight pods and turn on the headlights. Twilight Sentinel also delayed shutoff of headlights to allow exterior illumination after the ignition was turned off.

• Head Up Display (RPO UV6) projected instrument readouts onto the windshield to allow viewing without shifting vision to the instrument panel. The driver could view speedometer, tachometer, water temperature, oil pressure, fuel level and turn signals. Or, using a "page switch" feature, the driver could view just speed, just engine revolutions, or certain limited display combinations. A "check gauges" warning alerted the driver to check the instrument panel for a reading not displayed.

• Airbags for 1999 Corvettes were "next generation" meaning that they deployed with less force. A cutoff switch was not provided for the passenger side. But fifth-generation Corvettes were eligible for a dealer-installed cutoff switch kit for driver and/or passenger side. Kits included new seatbelts designed to work without airbag deployment.

• *Car and Driver* selected the 1999 Corvette as one of the 10-best cars (January 1999); *AutoWeek* readers picked it as America's best car (July 5, 1999).

1999 Options

RPO#	DESCRIPTION	QTY	RETAIL $
1YY07	Base Corvette Sport Coupe	18,078	$39,171.00
1YY37	Base Corvette Hardtop	4,031	38,777.00
1YY67	Base Corvette Convertible	11,161	45,579.00
AAB	Memory Package (coupe & conv)	23,829	150.00
AG1	Power Driver Seat (hardtop)	3,716	305.00
AG2	Power Passenger Seat (coupe & conv)	27,089	305.00
AQ9	Sport Seats (coupe & conv)	24,573	625.00
AP9	Parcel Net (hardtop)	2,738	15.00
B34	Floor Mats	32,706	25.00
B84	Body Side Moldings	19,348	75.00
C2L	Dual Removable Roof Panels (coupe)	6,307	950.00
CC3	Removable Roof Panel, blue tint (coupe)	5,235	650.00
CJ2	Dual Zone Air Conditioning	25,672	365.00
D42	Luggage Shade and Parcel Net (coupe)	18,058	50.00
F45	Selective Real Time Damping (coupe & conv)	7,515	1,695.00
G92	Performance Axle (3.15 ratio for automatic)	14,525	100.00
JL4	Active Handling System	20,174	500.00
MN6	6-Speed Manual Trans (coupe & conv)	13,729	825.00
N37	Telescopic Steering, Power (coupe & conv)	16,847	350.00
N73	Magnesium Wheels	2,029	3,000.00
T82	Twilight Sentinel (coupe & conv)	18,895	60.00
T96	Fog Lamps	28,546	69.00
TR9	Lighting Package (hardtop only)	3,037	95.00
UN0	Delco Stereo System with CD	20,442	100.00
UV6	Head Up Instrument Display	19,034	375.00
UZ6	Bose Speaker Package (hardtop)	3,348	820.00
U1S	Remote Compact 12-Disc Changer	16,997	600.00
V49	Front License Plate Frame	17,742	15.00
YF5	California Emissions	3,336	170.00
Z51	Performance Handling Pkg (coupe & conv)	10,244	350.00
86U	Magnetic Red Metallic Paint (coupe & conv)	2,733	500.00

• A 346ci, 345hp engine, Delco stereo cassette, and leather seats were included with all models.

• Coupe and convertible included 4-speed automatic transmission, Bose speakers, body-color roof panel (coupe), or soft top (convertible).

• Hardtop included Z51 suspension and 6-speed manual transmission.

• RPO AAB (Memory Package) remembered outside-rear-view-mirror, radio, hvac, and power driver seat settings. RPO CJ2 was required.

• RPO AQ9 (Sport Seats) required RPO AG2 (Power Passenger Seat).

• RPO C2L (Dual Removable Roof Panels) included standard body-color roof panel plus blue-tint transparent panel.

• RPOs CJ2, N73, T96 and UV6 were initially optional with coupe and convertible models only, but were available with hardtop models later in the production year.

• RPO F45 (Selective Real Time Damping) included driver-adjustable ride control system. Not available with RPO Z51.

• RPO Z51 (Performance Handling Package) included stiffer springs and stabilizers. Required RPO G92 with automatic; not available with F45.

1999 Colors

CODE	EXTERIOR	QTY	SOFT TOP	INTERIORS
10	Arctic White ✓	2,756	B-Lo-W	B-Lg-Lo-R
11	Light Pewter Metallic ✓	6,164	B-W	B-Lg-R
13	Sebring Silver Metallic	3,510	B-W	B-Lg-R
23	Nassau Blue Metallic ✓	1,034	B-Lo-W	B-Lg-Lo
28	Navy Blue Metallic	1,439	B-Lo-W	B-Lg-Lo
41	Black ✓	7,235	B-Lo-W	B-Lg-Lo-R
70	Torch Red ✓	8,361	B-Lo-W	B-Lg-Lo
86	Magnetic Red Metallic	2,733	B-Lo-W	B-Lg-Lo

✓ Available with hardtop model.

• Recommended color combinations shown; others could be ordered except hardtop model was limited to standard black interior.

• Coupe and convertible interior colors available in base or Sport.

Interior Codes: 193=B, 673=Lo, 923=Lg, 943=R.

Abbreviations: B-Black, Lg=Light Gray, Lo=Light Oak, R=Red, W=White

2000 Corvette

18,113 coupe, 13,479 convertible, 2,090 hardtop, 33,682 total.

2000 Numbers

Vehicle: 1G1YY22G6Y5100001 thru 1G1YY22G9Y5133610
- Sixth digit is a 1 for hardtop, 2 for coupe, 3 for convertible.
- Eighth digit is engine code: G=346ci, 345hp (LS1)
- Ninth digit is a security code and varies.
- Ending VIN did not match production quantity because 2000 production counted pilots and prototypes.

Suffix: ZBH: 346ci, 345hp, at ZBF: 346ci, 345hp, mt

Block: 12560626 & 12559378: 346ci, 345hp

Head: 12559853 & 12564241: 346ci, 345hp

Abbreviations: at=automatic transmission, ci=cubic inch, hp=horsepower, mt=manual transmission.

2000 Facts

• Corvette's wheels changed for 2000. Initially, a new thin-spoke design was standard. It was full-forged with a flow-formed rim which resulted in a more durable, "work hardened" rim section. This wheel was available with a polished finish as RPO QF5 at $895 additional cost. The popularity of the optional polished finish caused Chevrolet to announce in January 2000 that a new painted wheel from a different supplier would become standard to permit an increase in production of the polished wheel. This new standard wheel was similar in design, but with slightly thicker spokes. The design of the optional magnesium wheel, RPO N73, was unchanged but its cost was reduced from $3,000 to $2,000.

• The Performance Handling Package (RPO Z51) was upgraded with larger front and rear stabilizer bars, and revised shock absorber valving. These changes did not produce a noticeably stiffer ride, but did improve handling, especially in transient maneuvers, by increasing roll stiffness.

• Selective Real Time Damping (RPO F45) was refined using new or revised algorithms and a redesigned (softer) jounce bumper for ride and handling improvements.

• Two new exterior colors were available, Millennium Yellow and Dark Bowling Green Metallic. Like 1999's Magnetic Red Metallic exterior which carried over, Millennium Yellow cost an additional $500. This was not due to the paint itself, but to the special process and equipment required for a tinted clear coat which added visual depth to the paint's appearance.

• Torch Red, a new interior color, replaced Firethorn Red. Torch Red, also used in 1994-96 Corvettes, was a brighter hue and a better match for Torch Red exterior paint, typically Corvette's best seller.

• Passive Keyless Entry (PKE), pioneered by Corvette in its 1993 model, was replaced by conventional Active Keyless Entry. The passive system, which worked by proximity, was confusing to some owners.

• The passenger-side key cylinder lock was eliminated. It was deemed unnecessary with keyless entry, since the driver's side cylinder lock could still be used in case of battery failure.

• The tension of the spring that located the shifter between the first-second gear gate and the fifth-sixth gate in manual transmissions was increased to provide a better sense of gear location and selection (interim 1999).

• Two seals were redesigned to reduce water intrusion. The seal at the top of the windshield pillar was changed to minimize water entry when a door was opened. The coupe hatch seal was changed to better direct water away from the interior when the hatch was opened (interim 1999).

• Seat belt lap retractor snorkels were added to smooth lap belt webbing retraction. Also, the angle of the seat belt latch plates was changed to ease the buckling-in action.

• Dual Zone Air Conditioning calibrations were revised to improve comfort and defrosting, to clarify memory function, and to reduce noise.

• Seat materials and construction were changed to improve long-term durability. Some changes were phased in during 1999 production.

2000 Options

RPO#	DESCRIPTION	QTY	RETAIL $
1YY07	Base Corvette Sport Coupe	18,113	$39,475.00
1YY37	Base Corvette Hardtop	2,090	38,900.00
1YY67	Base Corvette Convertible	13,479	45,900.00
AAB	Memory Package (coupe & conv)	26,595	150.00
AG1	Power Driver Seat (hardtop)	1,841	305.00
AG2	Power Passenger Seat (coupe & conv)	29,462	305.00
AQ9	Sport Seats (coupe & conv)	27,103	700.00
AP9	Parcel Net (hardtop)	938	15.00
B34	Floor Mats	33,188	25.00
B84	Body Side Moldings	18,773	75.00
C2L	Dual Removable Roof Panels (coupe)	6,280	1,100.00
CC3	Removable Roof Panel, blue tint (coupe)	5,605	650.00
CJ2	Dual Zone Air Conditioning	29,428	365.00
D42	Luggage Shade and Parcel Net (coupe)	15,689	50.00
F45	Selective Real Time Damping (cpe & conv)	6,724	1,695.00
G92	Performance Axle (3.15 ratio for automatic)	14,090	100.00
JL4	Active Handling System	22,668	500.00
MN6	6-Speed Manual Trans (cpe & conv)	13,320	815.00
N37	Telescopic Steering, Power (cpe & conv)	22,182	350.00
N73	Magnesium Wheels	2,652	2,000.00
QF5	Polished Aluminum Wheels	15,204	895.00
T82	Twilight Sentinel (coupe & conv)	23,508	60.00
T96	Fog Lamps	31,992	69.00
TR9	Lighting Package (hardtop)	1,527	95.00
UN0	Delco Stereo System with CD	24,696	100.00
UV6	Head Up Instrument Display	26,482	375.00
UZ6	Bose Speaker Package (hardtop)	1,766	820.00
U1S	Remote Compact 12-Disc Changer	15,809	600.00
V49	Front License Plate Frame	17,380	15.00
YF5	California Emissions	3,628	0.00
Z51	Performance Handling (std w/hardtop)	7,775	350.00
79U	Millennium Yellow (coupe & conv)	3,578	500.00
86U	Magnetic Red Metallic Paint (cpe & conv)	2,941	500.00

• A 346ci, 345hp engine, Delco stereo cassette, and leather seats were included with all models.

• Coupe and convertible included 4-speed automatic transmission, Bose speakers, body-color roof panel (coupe), or soft top (convertible).

• Hardtop included Z51 suspension and 6-speed manual transmission.

• RPO AAB (Memory Package) remembered outside-rear-view-mirror, radio, hvac, and power driver seat settings. RPO CJ2 was required.

• RPO AQ9 (Sport Seats) required RPO AG2 (Power Passenger Seat).

• RPO C2L (Dual Removable Roof Panels) included standard body-color roof panel plus blue-tint transparent panel.

• RPO F45 (Selective Real Time Damping) included driver-adjustable ride control system. Not available with RPO Z51.

• RPO Z51 (Performance Handling Package) included stiffer springs and stabilizers. Required RPO G92 with automatic; not available with F45.

2000 Colors

CODE	EXTERIOR	QTY	SOFT TOP	INTERIORS
10	Arctic White ✓	1,979	B-Lo-W	B-Lg-Lo-R
11	Light Pewter Metallic ✓	5,125	B-W	B-Lg-R
13	Sebring Silver Metallic ✓	2,783	B-W	B-Lg-R
23	Nassau Blue Metallic ✓	851	B-Lo-W	B-Lg-Lo
28	Navy Blue Metallic	2,254	B-Lo-W	B-Lg-Lo
41	Black ✓	5,807	B-Lo-W	B-Lg-Lo-R
70	Torch Red ✓	6,700	B-Lo-W	B-Lg-Lo-R
79	Millennium Yellow	3,578	B-W	B
86	Magnetic Red Metallic	2,941	B-Lo-W	B-Lg-Lo
91	Dark Bowling Green Metallic	1,663	B-Lo-W	B-Lg-Lo

✓ Available with hardtop model.

• Other combos could be ordered, but hardtop had black std seats only.

• One 2000 Corvette coupe, VIN 1G1YY22G8Y5113591, was painted Platinum Purple Metallic and awarded to a Bowling Green employee.

Interior Codes: 193=B, 673=Lo, 703=R, 923=Lg.

Abbreviations: B-Black, Lg=Light Gray, Lo=Light Oak, R=Red, W=White

2001 Corvette

15,681 coupe, 14,173 conv, 5,773 Z06 hardtop, 35,627 total.

2001 Numbers

Vehicle: 1G1YY12S015100001 thru 1G1YY12S115135601
- Sixth digit is a 1 for hardtop, 2 for coupe, 3 for convertible.
- Eighth digit is engine code: G=346ci, 350hp (LS1)
                                  S=346ci, 385hp (LS6)
- Ninth digit is a security code and varies.
- Ending VIN did not match production quantity because 2001 production counted pilots and prototypes.

Suffix: ZCB: 346ci, 350hp, at   ZCA: 346ci, 385hp, mt
ZCC: 346ci, 350hp, mt

Block: 12561168: 346ci, 350hp, 385hp

Head: 12564241: 346ci, 350hp   12564243: 346ci, 385hp

Abbreviations: at=automatic transmission, ci=cubic inch, hp=horsepower, mt=manual transmission.

2001 Facts

- Using nomenclature from its past, Chevy introduced a Z06 model powered by a specific, more powerful engine, the 385hp-LS6. All 2001 hardtop models were Z06s with LS6 engines, and all had 6-speed manual transmissions with revised ratios.

- LS6 engines used a new block casting (shared with the base LS1 engine after production started), a modified air cleaner, new intake and exhaust manifolds (both shared with LS1), a unique camshaft profile, higher-rate valve springs, new cylinder heads, higher capacity fuel injectors, and a new PCV system that reduced crankcase pressure.

- Z06 models had titanium exhaust systems. McLaren F1s (107 built) had titanium exhausts, but Corvette's was the first such use for titanium in mass-production. Titanium saved 19.0 pounds compared to the stainless steel system found in Corvette's coupe and convertible models.

- Z06s included a transmission temperature sensor, and a unique wheel design with the same diameters as coupe and convertible, but one inch wider both front and rear. Goodyear Eagle F1 Supercar tires saved 23.4 pounds compared to the coupe and convertible tires. Unlike coupe and convertible, Z06's tires weren't run-flats, so a tire inflator kit was included since C5 Corvettes had no provision for spare tires. Z06s did not have a tire pressure monitor system.

- Z06's standard FE4 suspension had a larger, hollow front stabilizer bar, stiffer rear spring, and revised front and rear camber settings.

- Z06 interiors had leather seats with dual-density side bolster foam and Z06 embroidery on the headrests. Colors were black or a black/red combination. Instrument graphics had italicized numerals on checkered backgrounds; tachs had 6500rpm redlines compared to LS1's 6000rpm.

- Z06 exteriors had stainless steel mesh front fascia grilles, functional brake cooling ducts on the lower rear fenders, and red brake calipers. Exterior colors were limited to Speedway White, Quicksilver Metallic, Black, Torch Red, and extra-cost Millennium Yellow.

- LS1 horsepower (coupe and convertible) increased from 345hp to 350hp.

- Manual transmission synchronizers were upgraded with carbon blocker rings on all forward gears. Automatics were smoother at engagement due to a new alternator clutch pulley that allowed lower engine idle. Automatic transmission case weight was trimmed by 3.3 pounds.

- Maximum oil change interval increased from 10,000 to 15,000 miles.

- Second-generation active handling was standard with all 2001 models.

- All Corvette engines were produced at St. Catharines, Ontario, Canada.

- Convertible top noise was reduced with better weather stripping and a new, double-thickness twill material. Noise in all models was reduced by addition of an expandable foam baffle inside the lower lock pillar.

- Lighter, more durable Absorbent Glass Mat (AGM) batteries were standard with all 2001 Corvette models.

2001 Options

RPO#	DESCRIPTION	QTY	RETAIL $
1YY07	Base Corvette Sport Coupe	15,681	$40,475.00
1YY37	Base Corvette Z06 Hardtop	5,773	47,500.00
1YY67	Base Corvette Convertible	14,173	47,000.00
1SB	Preferred Equipment Group-Sport Coupe	2,514	1,639.00
1SB	Preferred Equipment Group-Convertible	1,710	1,769.00
1SC	Preferred Equipment Group-Sport Coupe	11,558	2,544.00
1SC	Preferred Equipment Group-Convertible	11,881	2,494.00
AAB	Memory Package (Z06 hardtop)	4,780	150.00
B34	Floor Mats	34,907	25.00
B84	Body Side Moldings	20,457	75.00
C2L	Dual Removable Roof Panels (coupe)	5,099	1,100.00
CC3	Removable Roof Panel, blue tint (coupe)	4,769	650.00
DD0	Electrochromic Mirrors (Z06 hardtop)	4,576	120.00
F45	Selective Real Time Damping (cpe & conv)	5,620	1,695.00
G92	Performance 3.15 Axle (automatic, cpe & conv)	12,882	300.00
MN6	6-Speed Manual Trans (cpe & conv)	16,019	815.00
N73	Magnesium Wheels (cpe & conv)	1,022	2,000.00
QF5	Polished Aluminum Wheels (cpe & conv)	22,980	895.00
R8C	Corvette Museum Delivery	457	490.00
UL0	Delco Stereo Cassette (replaces std radio)	6,844	-100.00
UN0	Delco Stereo System with CD	28,783	100.00
U1S	Remote 12-Disc Changer (cpe & conv)	14,198	600.00
V49	Front License Plate Frame	18,935	15.00
Z51	Performance Handling Pkg (cpe & conv)	7,817	350.00
79U	Millennium Yellow w/tint coat	3,887	600.00
86U	Magnetic Red Metallic Paint (cpe & conv)	3,322	600.00

• All models included Delco stereo CD, leather seats, 6-way power driver seat, and active handling system.

• Coupe and convertible models included 346ci, 350hp engine, automatic transmission, air conditioning, and tire-pressure monitoring system.

• Z06 hardtop model included 346ci, 385hp engine, 6-speed manual transmission, electronic dual-zone air conditioning, and FE4 suspension.

• Some options included in "preferred equipment groups" couldn't be ordered separately, but this varied. For example, RPO DD0 was available separately for Z06 hardtops, but package-only for coupe and convertible.

• 1SB coupe equipment group included sport bucket seats, 6-way power passenger seat, electronic dual-zone air conditioning, fog lamps, memory package, and luggage/parcel shade.

• 1SB convertible equipment group included sport bucket seats, 6-way power passenger seat, electronic dual-zone air conditioning, fog lamps, memory package, twilight sentinel and electrochromic mirrors.

• 1SC coupe equipment group included 1SB equipment plus electrochromic mirrors, head-up instrument display, twilight sentinel and power-telescopic steering column.

• 1SC convertible equipment group included 1SB equipment plus head-up instrument display and power-telescopic steering column.

• RPO C2L (Dual Removable Roof Panels) included standard body-color roof panel plus blue-tint transparent panel.

• Base prices, groups and some options increased in cost during 2001.

2001 Colors

		Quantity				
Code	Exterior	Cpe	Z06	Conv	Total	Interior
11	Light Pewter	1,983	na	1,479	3,462	B-Lg-R
12	Quicksilver	1,905	1,345	1,572	4,822	B-Lg-R
28	Navy Blue	1,388	na	1,199	2,587	B-Lg-Lo
40	Speedway White	1,066	352	1,047	2,465	B-Lg-Lo-R
41	Black	2,450	1,819	2,702	6,971	B-Lg-Lo-R
70	Torch Red	2,788	1,623	2,781	7,192	B-Lg-Lo-R
79	Millennium Yellow	1,822	634	1,431	3,887	B
86	Magnetic Red	1,836	na	1,486	3,322	B-Lg-Lo
91	Dark Bowling Green	443	na	476	919	B-Lg-Lo

• Recommended color combinations shown; others could be ordered, but unique Z06 hardtop interiors were restricted to black or black/red.

• Coupe and convertible interior colors available in base or Sport.

Interior Codes: 193=B, 194=B (Z06), 673=Lo, 703=R, 704=B/R (Z06), 923=Lg.

Abbreviations: B-Black, Lg=Light Gray, Lo=Light Oak, na=not available, R=Red, W=White

2002 Corvette

2002 Numbers

Vehicle: 1G1YY22G725100001 thru 1G1YY22G125135763
- Sixth digit is a 1 for hardtop, 2 for coupe, 3 for convertible.
- Eighth digit is engine code: G=346ci, 350hp (LS1)
                                    S=346ci, 405hp (LS6)
- Ninth digit is a security code and varies.
- Ending VIN did not match production quantity due to pilots, prototypes and two VINs (1183 &1939) not built.

Suffix: ZDD: 346ci, 350hp, at   ZDH: 346ci, 405hp, mt
ZDF: 346ci, 350hp, mt

Block: 12561168: 346ci, 350hp, 405hp

Head: 12564241: 346ci, 350hp   12564243: 346ci, 405hp

Abbreviations: at=automatic transmission, ci=cubic inch, hp=horsepower, mt=manual transmission.

2002 Facts

• Automatic transmission cooler cases changed from stainless steel to aluminum. RPO N73 magnesium wheels weren't listed as optional, but were available at $1,500 in limited quantities. Floor mats were standard.

• Two exterior colors, Dark Bowling Green Metallic and Navy Blue Metallic were deleted; Electron Blue Metallic was new. Coupes and convertibles were available in all eight exterior colors. Z06 hardtop models were available in five colors: Electron Blue Metallic, Quicksilver Metallic, Black, Torch Red and extra-cost Millennium Yellow.

• Front stabilizer links were changed from rolled rod steel to aluminum for weight savings. Intended for Z51 or FE4 (Z06) suspensions, the links were used on all 2002 models for assembly consistency.

• Convertibles equipped with head-up display and all Z06s had a thinner 4.8 mm windshield compared to the standard 5.4 mm, a weight savings of 2.65 pounds per car. Convertibles without head-up display and all coupes continued to have the thicker windshield.

• The LS6 engine standard with Z06 hardtop models increased in output from 385 to 405 horsepower. Torque increased from 385 to 400 lb-ft. Front fender badges for Z06 models included the 405 HP designation.

• LS6 exhaust back pressure was reduced 16% by the removal of two small catalytic converters in the takedown pipes, leaving only the larger, primary under-floor converters. Though the shape and size of the two larger converters remained the same, their internal composition of palladium and rhodium was altered so that Nationwide Low Emissions Vehicle standards could be met. The camshaft was redesigned with the highest lift in small-block Chevy V8 history. The increased lift resulted in a 5% increase in air flow through the cylinder chamber. Valve spring material composition was changed to increase seat load by 14%.

• LS6 intake and exhaust valves were a new, lighter design. Intake valves were hollow. Exhaust valves, which operate at higher temperature, were hollow but filled with a liquid alloy of 78% potassium and 22% sodium for better heat transfer. The valvetrain's weight was reduced by 0.81 pounds, which improved valve seating at high rpm.

• Airflow over the LS6's mass airflow (MAF) sensor was increased by removal of gridwork that had been used to smooth air flow. MAF softwear was recalibrated to read air flow without the grid. Also, LS6's air cleaner housing had a larger intake to increase air flow about 5%. The air cleaner filter element itself was unchanged.

• Appearance of Z06 aluminum wheels was the same, but they were "cast spun" instead of forged for weight savings of 1.3 pounds per car.

• Other Z06 upgrades included revised rear shock absorber valving for better traction and smoother ride over rough surfaces, improved clutch, thicker front anti-sway bar, and a modified steering rack that reduced turning radius by about two feet. Z06 and export models also had new front brake pad composition for reduced fade and increased durability.

2002 Options

RPO#	DESCRIPTION	QTY	RETAIL $
1YY07	Base Corvette Sport Coupe	14,760	41,450.00
1YY37	Base Corvette Z06 Hardtop	8,297	50,150.00
1YY67	Base Corvette Convertible	12,710	47,975.00
1SB	Preferred Equipment Group-Sport Coupe	1,359	1,700.00
1SB	Preferred Equipment Group-Convertible	1,379	1,800.00
1SC	Preferred Equipment Group-Sport Coupe	11,136	2,700.00
1SC	Preferred Equipment Group-Convertible	10,964	2,600.00
AAB	Memory Package (Z06 hardtop)	7,794	150.00
B84	Body Side Moldings	21,422	75.00
C2L	Dual Removable Roof Panels (coupe)	5,079	1,200.00
CC3	Removable Roof Panel, blue tint (coupe)	4,208	750.00
DD0	Electrochromic Mirrors (Z06 hardtop)	7,394	120.00
F45	Selective Real Time Damping (cpe & conv)	4,773	1,695.00
G92	Performance 3.15 Axle (automatic, cpe & conv)	9,646	300.00
MN6	6-Speed Manual Transmission (cpe & conv)	8,553	815.00
N73	Magnesium Wheels (cpe & conv)	114	1,500.00
QF5	Polished Aluminum Wheels (cpe & conv)	22,597	1,200.00
R8C	Corvette Museum Delivery	371	490.00
UL0	Delco Stereo Cassette (cpe & conv)	4,210	-100.00
U1S	Remote 12-Disc Changer (cpe & conv)	13,725	600.00
V49	Front License Plate Frame	19,948	15.00
Z51	Performance Handling Pkg (cpe & conv)	6,106	350.00
79U	Millennium Yellow w/tint coat	4,040	600.00
86U	Magnetic Red Metallic Paint (cpe & conv)	3,298	600.00

• All models included Delco stereo CD, leather seats, 6-way power driver seat, active handling system, and floor mats.

• Coupe and convertible models included 346ci, 350hp engine, automatic transmission, air conditioning, and tire-pressure monitoring system.

• Z06 hardtop included 346ci, 405hp engine, , FE4 suspension, 6-speed manual transmission, electronic dual-zone a/c, and head-up display.

• Some previously stand-alone options were included in "preferred equipment groups" and couldn't be ordered separately. This varied by model. For example, RPO DD0 was available separately for Z06 hardtop models, but package-only for coupe and convertible.

• 1SB coupe equipment group included sport bucket seats, 6-way power passenger seat, electronic dual-zone air conditioning, fog lamps, memory package, and luggage/parcel shade.

• 1SB convertible equipment group included sport bucket seats, 6-way power passenger seat, electronic dual-zone air conditioning, fog lamps, memory package, twilight sentinel and electrochromic mirrors.

• 1SC coupe equipment group included 1SB equipment plus electrochromic mirrors, head-up instrument display, twilight sentinel and power-telescopic steering column.

• 1SC convertible equipment group included 1SB equipment plus head-up instrument display and power-telescopic steering column.

• RPO C2L (Dual Removable Roof Panels) included standard body-color roof panel plus blue-tint transparent panel.

• RPO UL0 Stereo Cassette was available only when combined with Remote Compact 12 Disc Changer (RPO U1S).

2002 Colors

		Quantity				
Code	Exterior	Cpe	Z06	Conv	Total	Interior
11	Light Pewter	1,578	na	1,072	2,650	B-Lg-R
12	Quicksilver	1,808	1,467	1,343	4,618	B-Lg-R
21	Electron Blue	2,222	1,814	1,371	5,407	B-Lg-Lo
40	Speedway White	879	na	884	1,763	B-Lg-Lo-R
41	Black	2,522	2,205	2,402	7,129	B-Lg-Lo-R
70	Torch Red	2,830	1,746	2,286	6,862	B-Lg-Lo-R
79	Millennium Yellow	1,312	1,065	1,663	4,040	B
86	Magnetic Red	1,609	na	1,689	3,298	B-Lg-Lo

• Recommended color combinations shown; others could be ordered, but unique Z06 hardtop interiors were restricted to black or black/red.

• Coupe and convertible interior colors available in base or Sport.

Interior Codes: 193=B, 194=B (Z06), 673=Lo, 703=R, 704=B/R (Z06), 923=Lg.

Abbreviations: B-Black, Lg=Light Gray, Lo=Light Oak, na=not available, R=Red, W=White

2003 Corvette

12,812 coupe, 14,022 conv, 8,635 Z06 hardtop, 35,469 total.

2003 Numbers

Vehicle: 1G1YY22G535100001 thru 1G1YY32G235135469
- Sixth digit is a 1 for hardtop, 2 for coupe, 3 for convertible.
- Eighth digit is engine code: G=346ci, 350hp (LS1)
                                                    S=346ci, 405hp (LS6)
- Ninth digit is a security code and varies.

Suffix: ZFA: 346ci, 350hp, at   ZFB: 346ci, 405hp, mt
                    ZFC: 346ci, 350hp, mt

Block: 12561168: 346ci, 350hp, 405hp

Head: 12564241: 346ci, 350hp   12564243: 346ci, 405hp

Abbreviations: at=automatic transmission, ci=cubic inch, hp=horsepower, mt=manual transmission.

2003 Facts

• It had been fifty years since the Corvette's 1953 introduction and the occasion was marked with an optional 50th Anniversary package. Ordered as the 1SC equipment group, it was available with coupes and convertibles, but not Z06 hardtops. It included Anniversary Red "Xirallic crystal" exterior paint, unique front-fender emblems, champagne-painted aluminum wheels and center caps, and a new Shale interior. These models had 1SB content along with a new F55 Magnetic Selective Ride Control system which was optional on standard coupes and convertibles.

•The 50th Anniversary's Shale interior blended a lighter value of gray-beige for seats and carpeting with a darker gray-beige for console, instrument panel and upper door panels. This was the first appearance of anything but black for a C5 instrument panel. The 50th Anniversary interior included embroidered logos on the seat headrests and floor mats, and leather-padded door armrests and inside door pull handles. 50th Anniversary convertibles also had Shale color soft tops.

• RPO F55 Magnetic Selective Ride Control replaced the earlier F45 Real Time Damping system. It featured faster response time made possible by the use of magnetic fluid within the shock absorbers. The synthetic fluid, called Magneto-rheological (MR), held iron particles in suspension. Electrical input caused the iron particles to align in a way that effected flow rate through orifices in the monotube shock's piston. In the earlier system (and virtually all competing systems at the time), shock damping adjustments were made by mechanically varying the orifice size of valves through which fluid flowed. Corvette's new magnetic system adjusted shock damping at the rate of 1,000 times per second. This equated to the ability of reacting to every inch of road surface at 60 mph.

• Several features previously optional or included in option packages were standard equipment in 2003 coupes and convertibles. The list included foglamps, sport seats, power passenger seat, dual-zone air conditioning, and (for coupe only) the parcel net and luggage shade.

• Exterior colors Light Pewter Metallic and Magnetic Red Metallic were deleted. Medium Spiral Gray Metallic and 50th Anniversary Red Metallic were new, the latter exclusive to 50th Anniversary trim models.

• To meet stricter standards for occupant protection, Z06 hardtop headliners were thicker. Also, the interior areas at the A-pillars for all models and the B-pillars for coupes and Z06 hardtops were modified.

• The Indianapolis 500 race on May 26, 2002, was paced by Corvette, the fifth time Corvette had been so honored. A special Corvette pace model was not built; rather a nearly-stock 2003 50th Anniversary package coupe was used. An Indy 500 graphics package was available for $495.

• All models featured LATCH (lower assist tether for children) child seat hooks. Child seat use required the passenger airbag to be switched off.

• All 2003s had 50th anniversary emblems on their hoods and rear decks, and on nomenclature such as manuals and key blanks. Coupe and convertible tachometer and speedometer displays also had 50th emblems.

2003 Options

RPO#	DESCRIPTION	QTY	RETAIL $
1YY07	Base Corvette Sport Coupe	12,812	$43,895.00
1YY37	Base Corvette Z06 Hardtop	8,635	51,155.00
1YY67	Base Corvette Convertible	14,022	50,370.00
1SB	Preferred Equipment Group-Sport Coupe	7,310	1,200.00
1SB	Preferred Equipment Group-Convertible	6,643	1,200.00
1SC	50th Anniversary Edition-Sport Coupe	4,085	5,000.00
1SC	50th Anniversary Edition-Convertible	7,547	5,000.00
AAB	Memory Package (Z06 hardtop)	8,241	175.00
B84	Body Side Moldings	22,243	150.00
C2L	Dual Removable Roof Panels (coupe)	5,184	1,200.00
CC3	Removable Roof Panel, blue tint (coupe)	3,150	750.00
DD0	Electrochromic Mirrors (Z06 hardtop)	8,227	120.00
F55	Magnetic Selective Ride Control (cpe & conv)	14,992	1,695.00
G92	Performance 3.15 Axle (auto, cpe & conv)	9,785	395.00
MN6	6-Speed Manual Transmission (cpe & conv)	8,590	915.00
N73	Magnesium Wheels (cpe & conv)	293	1,500.00
QF5	Polished Aluminum Wheels (cpe & conv)	10,290	1,295.00
R8C	Corvette Museum Delivery	787	490.00
UL0	Delco Stereo Cassette (cpe & conv)	4,664	0.00
U1S	Remote 12-Disc Changer (cpe & conv)	15,979	600.00
V49	Front License Plate Frame	20,605	15.00
Z51	Performance Handling Package (cpe & conv)	2,592	395.00
79U	Millennium Yellow w/tint coat	3,900	750.00

• All models included Delco stereo CD, leather seats, 6-way power driver seat, active handling system, dual-zone electronic air conditioning, and floor mats.

• Coupe and convertible models included 346ci, 350hp engine, automatic transmission, sport seats, 6-way power passenger seat, foglamps, and tire-pressure monitoring system.

• Z06 hardtop model included 346ci, 405hp engine, 6-speed manual transmission, FE4 suspension, and head-up display.

• RPOs AAB and DD0 were stand-alone options for Z06 hardtop models, but package-only for coupe and convertible.

• 1SB coupe equipment group included head-up display, power-telescopic steering column, electrochromic mirrors, memory package and twilight sentinel.

• 1SB convertible equipment group included head-up display, power-telescopic steering column, electrochromic mirrors, memory package and twilight sentinel.

• 1SC coupe equipment group included 1SB equipment plus Magnetic Selective Ride Control, and 50th anniversary paint and trim.

• 1SC convertible equipment group included 1SB equipment plus Magnetic Selective Ride Control, and 50th anniversary paint and trim.

• RPO C2L (Dual Removable Roof Panels) included standard body-color roof panel plus blue-tint transparent panel.

• RPO UL0 Stereo Cassette was available only when combined with Remote Compact 12 Disc Changer (RPO U1S).

2003 Colors

Code	Exterior	Cpe	Z06	Conv	Total	Interior
12	Quicksilver	798	1,089	500	2,387	B-Lg-R
21	Electron Blue	967	1,272	531	2,770	B-Lg-Lo
40	Speedway White	371	na	308	679	B-Lg-Lo-R
41	Black	1,600	2,599	1,398	5,597	B-Lg-Lo-R
70	Torch Red	1,864	1,967	1,374	5,205	B-Lg-Lo-R
79	Millennium Yellow	1,041	1,708	1,151	3,900	B
88	Medium Spiral Gray	2,086	na	1,213	3,299	B-Lg-Lo-R
94	Anniversary Red	4,085	na	7,547	11,632	S

• Recommended combinations shown; others could be ordered, except Z06 interiors restricted to black or black/red, and the 50th Anniversary package interior was shale only. Soft tops available in black, light oak and white except Anniversary Red which was shale only.

Interior Codes: 154=S, 193=B, 194=B (Z06), 673=Lo, 703=R, 704=B/R (Z06), 923=Lg.

Abbreviations: B-Black, Lg=Light Gray, Lo=Light Oak, na=not available, R=Red, S=Shale, na=not available.

2004 Corvette
16,165 coupe, 12,216 conv, 5,683 Z06 hardtop, 34,064 total.

2004 Numbers

Vehicle: 1G1YY22G345100001 thru 1G1YY22GX45134064
- Sixth digit is a 1 for hardtop, 2 for coupe, 3 for convertible.
- Eighth digit is engine code: G=346ci, 350hp (LS1)
  - S=346ci, 405hp (LS6)
- Ninth digit is a security code and varies.

Suffix: ZHH: 346ci, 350hp, at   ZFL: 346ci, 350hp, at, eee
ZHJ: 346ci, 350hp, mt   ZHM: 346ci, 350hp, mt, eee
ZHK: 346ci, 405hp, mt

Block: 12561168: 346ci, 350hp, 405hp

Head: 12564241: 346ci, 350hp   12564243: 346ci, 405hp

Abbreviations: at=automatic transmission, ci=cubic inch, eee=early european export, hp=horsepower, mt=manual transmission.

2004 Facts

• Commemorative editions for coupe, convertible and Z06 hardtop models were offered to honor the success of the Corvette C5R racer, and to close out this last year of the C5 generation. All commemorative editions had LeMans Blue exterior paint, crossed-flag embroidery on seat headrests, unique commemorative road wheel centers, and silver front and rear emblems with inserts to recall Corvette's success at LeMans with the words "Commemorative 24:00 Heures Du Mans 2 GTS Wins."

• Commemorative coupes and convertibles had "Commemorative" stitched into their seat backs, and convertibles had silver crossed flag commemorative emblems on their between-seatback "waterfall" panels.

• Commemorative Z06s had features not found on convertibles and coupes. Most significant was a carbon fiber hood. Weighing just 20.5 lbs (compared to 31.1 for a conventional hood), this was the first use of carbon fiber for a painted exterior panel in a North American production vehicle. Z06s had a C5R-inspired silver and red striping scheme for the hood, roof and rear deck. Also, Commemorative Z06 wheels, while the same design as standard Z06s, were polished, a first for C5 Z06 models.

• Carbon fibers in the Z06's hood were aligned in a single direction to improve painted appearance. A cross-hatch pattern was subtly displayed in the hood graphic between the silver and red sections. The hood's carbon fiber technology was aerospace based, and was a collaboration between Chevrolet, MacLean Systems, and Toray Composites. The hood's liner was a hybrid of carbon fiber and Sheet Molded Compound (SMC).

• Coupe and convertible Commemorative Editions had Shale interiors which blended a lighter value of gray-beige for seats and carpeting with a darker gray-beige for console, instrument panel and upper door panels. This interior was similar, but not identical, to that included with 2003's Anniversary models. Z06 Commemorative Editions had black interiors.

• Z06 suspensions were revised to improve cornering control without ride quality sacrifice. Changes included refined shock absorber valving, stiffer upper control-arm bushings, and softer rear anti-roll bar bushings. Perfected at Germany's Nurburgring racetrack, the 2004 Z06 was able to break the 8-minute-per-lap barrier, an unofficial benchmark for supercars.

• The Z06 was not exported to Europe, but Europeans could get a special LeMans Blue Commemorative coupe (Z18) with Z06 stripes, suspension, black interior, and carbon fiber hood. It had non-EMT tires in the stock coupe size, and a 6-speed manual with the LS1. Total sales were 46.

• Arctic White, a brighter shade that had been available with C5s from 1997 to 2000, replaced Speedway White. Magnetic Red II was new, a reformulated version of the Magnetic Red available from 1999 to 2002. Machine Silver replaced Quicksilver. Anniversary Red was discontinued.

• The 2004 Cadillac XLR, a luxury two-seater with a power-retractable hardtop, was built on a separate line at Corvette's Bowling Green facility.

• A 2004 Corvette convertible paced the May 30th, 2004 Indy 500.

2004 Options

RPO#	DESCRIPTION	QTY	RETAIL $
1YY07	Base Corvette Sport Coupe	16,165	$44,535.00
1YY37	Base Corvette Z06 Hardtop	5,683	52,385.00
1YY67	Base Corvette Convertible	12,216	51,535.00
1SB	Preferred Equipment Group-Sport Coupe	11,446	1,200.00
1SB	Preferred Equipment Group-Convertible	9,334	1,200.00
1SB	Commemorative Edition-Z06	2,025	4,335.00
1SC	Commemorative Edition-Sport Coupe	2,215	3,700.00
1SC	Commemorative Edition -Convertible	2,659	3,700.00
AAB	Memory Package (Z06 hardtop)	5,446	175.00
B84	Body Side Moldings	20,626	150.00
C2L	Dual Removable Roof Panels (coupe)	5,079	1,400.00
CC3	Removable Roof Panel, blue tint (coupe)	4,356	750.00
DD0	Auto-dimming Mirrors (Z06 hardtop)	5,446	160.00
F55	Magnetic Selective Ride Control (cpe & conv)	5,843	1,695.00
G92	Performance 3.15 Axle (auto, cpe & conv)	10,367	395.00
MN6	6-Speed Manual Transmission (cpe & conv)	6,928	915.00
N73	Magnesium Wheels (cpe & conv)	1,110	995.00
QF5	Polished Aluminum Wheels (cpe & conv)	22,487	1,295.00
R8C	Corvette Museum Delivery	142	490.00
UL0	Delco Stereo Cassette (cpe & conv)	3,860	0.00
U1S	Remote 12-Disc Changer (cpe & conv)	14,668	600.00
V49	Front License Plate Frame	19,520	15.00
Z51	Performance Handling Package (cpe & conv)	3,672	395.00
79U	Millennium Yellow w/tint coat	2,641	750.00
86U	Magnetic Red II	3,596	750.00

• All models included Delco stereo CD, leather seats, 6-way power driver seat, active handling system, traction control, dual-zone electronic air conditioning, tilt wheel, cruise control, foglamps, and floor mats.

• Coupe and convertible models included 346ci, 350hp LS1 engine, automatic transmission, sport seats, 6-way power passenger seat, foglamps, and tire-pressure monitoring system.

• Z06 hardtop model included 346ci, 405hp LS6 engine, 6-speed manual transmission, FE4 suspension, and head-up display.

• RPOs AAB and DD0 were available optionally in combination for Z06 hardtop models, but package-only for coupe and convertible.

• 1SB coupe and convertible equipment group included head-up display, power-telescopic steering column, auto-dimming mirrors, memory package and twilight sentinel.

• 1SB Z06 equipment included Commemorative Edition LeMans Blue exterior paint and trim, black interior, silver and red striping for hood, roof and rear deck, carbon fiber hood, and polished Z06-style wheels.

• 1SC coupe and convertible equipment group included 1SB equipment plus Magnetic Selective Ride Control, Commemorative Edition LeMans blue exterior paint and trim, Shale interior, and polished (QF5) wheels.

• RPO C2L (Dual Removable Roof Panels) included standard body-color roof panel plus blue-tint transparent panel.

2004 Colors

Code	Exterior	Cpe	Z06	Conv	Total	Interior
10	Arctic White	921	na	820	1,741	B-Lg--Lo-R
19	Lemans Blue (ComEd)	2,215	2,025	2,659	6,899	S-B
41	Black	3,105	1,186	1,921	6,212	B-Lg-Lo-R
67	Machine Silver	2,514	967	1,185	4,666	B-Lg-R
70	Torch Red	2,438	876	1,709	5,023	B-Lg-Lo-R
79	Millennium Yellow	918	629	1,094	2,641	B
86	Magnetic Red II	1,948	na	1,648	3,596	B-Lg-Lo
88	Medium Spiral Gray	2,106	na	1,180	3,286	B-Lg-Lo-R

(Quantity column headers: Cpe, Z06, Conv, Total)

• Recommended combinations shown. Others could be ordered, except Commemorative coupes and convertibles had shale interiors only, Commemorative Z06s had black interiors, standard Z06 interiors were restricted to black or black/red. Soft tops were available in black, light oak and white, except Commemorative convertible was shale only.

• Interior quantities were: 20,979 Black, 4,829 Shale, 3,983 Lt Oak, 1,535 Red, and 2,738 Lt Gray. (Euro Z18 coupe had code 19 with black interior.)

Interior Codes: 152=S (coupe/conv commemorative), 192=B (Z06 commemorative), 193=B, 194=B (Z06), 673=Lo, 703=R, 704=B/R (Z06), 923=Lg.

Abbreviations: B-Black, Lg=Light Gray, Lo=Light Oak, R=Red, S=Shale.

2005 Corvette

26,728 coupe, 10,644 conv, 37,372 total.

2005 Numbers

Vehicle: 1G1YY24U855100001 thru 1G1YY24U855137341
- Sixth digit is a 2 for coupe, 3 for convertible.
- Seventh digit is restraint code: 2=front, 4=front/side
- Eighth digit is engine code: U=364ci, 400hp (LS2)
- Ninth digit is a security code and varies.
- Ending VIN did not match total quantity because 2005 production counted 31 prototype builds.

Suffix: ZJA: 364ci, 400hp, at   ZJB: 364ci, 400hp, mt

Block: 12568952: 364ci, 400hp

Head: 12564243: 364ci, 400hp

Abbreviations: at=automatic transmission, ci=cubic inch, hp=horsepower, mt=manual transmission.

2005 Facts

• First of the Corvette's sixth generation, the 2005 retained C5's front-engine, rear-transmission layout, but its body and interior were new, most chassis parts were re-engineered, and nearly all dimensions changed. Length was reduced 5.1 inches and width 1.1 inches. Wheelbase increased 1.2 inches to 105.7. The increased wheelbase and efficient packaging permitted interior and storage volumes to remain similar.

• 2005 models had exposed headlights, not seen in Corvettes since the 1962 model. 2005s used Xenon high intensity discharge low-beam projector lenses, and tungsten-halogen high-beam projectors under clear polycarbonate covers. Also under the covers were parking lights, side-turn markers and daytime running lights, all set into body-color housings.

• New options included DVD-based navigation, OnStar, XM Satellite radio, heated seats and seat-mounted side impact air bags. A power top for convertibles, last available with 1962 Corvettes, was optional.

• The 2005 coupe was shown at the Detroit International Auto Show in January 2004; the convertible at Geneva's Auto Show in March 2004.

• The standard engine for 2005 models was a 4th generation Chevy small block, the 400-hp LS2. Displacement increased from 346ci (5.7L) to 364ci (6.0L) by enlarging the bore from 3.9 inches to 4.0. Torque increased to 400 lb-ft. The LS2 had numerous refinements, including increased camshaft lift, a higher compression ratio (10.9:1), lighter exhaust manifolds and a wingless aluminum oil pan that improved oil control during high g-force driving while reducing oil capacity from 6.5 to 5.5 quarts with filter. Redline for the LS2 was 6500rpm, compared to LS1's 6000rpm.

• Six-speed manual transmissions were extra cost for C5, but included in 2005's base price. A four-speed automatic was a no-cost option. The manual featured a tighter shift pattern and a 1-inch shorter shift lever. It required shift to reverse at engine shutdown. The automatic was the 4L65-E, a stronger version of C5's 4L60-E. It had an electronic controller with performance shift algorithm which adapted to driver patterns.

• Wheel size was 18x8.5" front, 19x10" rear. Tire size was P245/40ZR-18 front; P285/35ZR-19 rear. Standard tires were Goodyear Eagle F1 EMT.

• The Z51 Performance Package initial option cost was $1,495 compared to 2004's $395, reflecting more content. The 2005 Z51 package included higher spring rates and shocks, larger sway bars, larger cross-drilled brake rotors; engine oil, transmission and power steering fluid coolers; higher-grip asymmetrical Goodyear Supercar EMT tires, and revised gear ratios with 6-speed manual, or performance axle ratio (3.15:1) with automatic.

• The 2005 had keyless access and start. A standard key was included for glovebox, console and emergency entry, but normal door lock/unlock and push-button starting were activated by keyfob proximity.

• A solid, bright red exterior was not available initially. The Bowling Green plant built one Torch Red 2005 to demonstrate its viability and Precision Red was replaced by the brighter Victory Red. The one Torch Red 2005 was acquired and sold by the National Corvette Museum.

• A red 2005 Corvette convertible paced the May 29th, 2005 Indy 500.

2005 Options

RPO#	DESCRIPTION	QTY	RETAIL $
1YY07	Base Corvette Coupe	26,728	$44,245.00
1YY67	Base Corvette Convertible	10,644	52,245.00
1SA	Preferred Equipment Group-Coupe	3,763	1,405.00
1SB	Preferred Equipment Group-Coupe	22,319	4,360.00
1SB	Preferred Equipment Group-Conv	10,306	2,955.00
C2L	Dual Removable Roof Panels (coupe)	2,585	1,400.00
CC3	Removable Roof Panel, transparent (cpe)	8,469	750.00
CM7	Power Convertible Top	7,541	1,995.00
F55	Magnetic Selective Ride Control	9,041	1,695.00
G90	Performance 3.15 Axle (w/auto trans)	15,112	395.00
MX0	Four-speed Automatic Transmission	22,380	0.00
QG7	Polished Aluminum Wheels	27,080	1,295.00
QX1	Competition Gray Alum Wheels (late)	621	295.00
R8C	Corvette Museum Delivery	831	490.00
UE1	OnStar System	19,634	695.00
U2K	XM Satellite Radio	21,896	325.00
U3U	AM/FM CD w/DVD Nav, Bose	4,676	1,400.00
Z51	Performance Package	15,345	1,495.00
19U	LeMans Blue exterior paint	3,759	300.00
45U	Velocity Yellow exterior paint (late)	760	750.00
79U	Millennium Yellow exterior paint	2,002	750.00
80U	Monterey Red exterior paint (late)	717	750.00
86U	Magnetic Red exterior paint	3,404	750.00

• All models included 364ci (6.0L), 400hp V8 engine, 6-speed manual or 4-speed automatic transmission, stereo CD with MP3 capability, leather seats, 6-way power driver seat, active handling, traction control, dual-zone electronic air conditioning, keyless access and start, foglamps, tilt steering wheel, cruise control, tire pressure monitor, and floor mats.

• Coupe had body-color removable roof panel, power hatch release with power pulldown. Convertible had power trunk release and manual soft top.

• 1SA coupe equipment and base convertible models included seat-mounted side impact air bags, sport seats with perforated leather, 6-way power passenger seat, cargo net, and luggage shade (coupe).

• 1SB coupe and convertible equipment included 1SA equipment plus head-up display, power-telescopic steering column, auto-dimming inside and outside mirrors, memory package, heated seats, premium Bose stereo with 6-disc in-dash CD changer, and homelink (garage door) transmitter.

• Power convertible top (CM7), OnStar (UE1), XM Satellite Radio (U2K), and Navigation (U3U) were optional only with 1SB equipment groups.

• The U3U CD Navigation system included a 6.5" LCD color touch screen with voice recognition. U3U deleted the 6-disc in-dash CD and MP3.

• Magnetic Selective Ride Control (F55) was available with coupe or convertible; not available with Z51 Performance Package.

• Performance Package (Z51) included specific springs, shocks and stabilizer bars, larger cross-drilled brake rotors, engine oil, transmission and power steering fluid coolers, Goodyear Supercar EMT asymmetrical tires. Also, quicker gear ratios (6-speed) or G90 3.15:1 axle (automatic).

2005 Colors

Code	Exterior	Coupe	Conv	Total	Interior
10	Arctic White	1,472	496	1,968	R-E-C-SG
19	LeMans Blue	2,718	1,041	3,759	E-C-SG
27	Precision Red	1,273	30	1,303	R-E-C-SG
27*	Torch Red	1	0	1	R-E-C-SG
41	Black	5,467	2,528	7,995	R-E-C-SG
45	Velocity Yellow	359	401	760	E-C-SG
67	Machine Silver	4,789	2,076	6,865	R-E-C-SG
71	Daytona Sunset Orange	2,379	602	2,981	E-C
74	Victory Red	3,285	2,332	5,617	R-E-C-SG
79	Millennium Yellow	1,646	356	2,002	E-C
80	Monterey Red	358	359	717	R-E-C-SG
86	Magnetic Red	2,981	423	3,404	R-E-C-SG

• Required combinations shown. Base seats were available only in Ebony.

• Precision Red available early, then replaced by Victory Red. Velocity Yellow and Monterey Red were 2006 colors available for late 2005 models.

* One factory-painted Torch Red 2005 had Precision Red's 27 paint code.

Interior Codes: 023=Red, 193=Ebony, 313=Cashmere, 363=Steel Gray.

Abbreviations: B=Black, Be=Beige, R=Red, E=Ebony, C=Cashmere, SG=Steel Gray.

2006 Corvette

16,598 coupe, 11,151 conv, 6,272 Z06, 34,021 total.

2006 Numbers

Vehicle: 1G1YY26UX65100001 thru 1G1YY26E865133992
- Sixth digit is a 2 for coupe and Z06, 3 for convertible.
- Seventh digit is restraint code: 4=front, 6=front/side
- Eighth digit is engine code: U=364ci, 400hp (LS2); E=427ci, 505hp (LS7), except early Z06 to VIN 100162 had Y engine code.
- Ninth digit is a security code and varies.
- Ending VIN did not match total quantity because 2006 production counted 29 non-salable prototype builds.

Suffix: ZKS: 364ci, 400hp, at   ZKH: 364ci, 400hp, mt
ZKD: 427ci, 505hp, mt

Block: 12568952: 364ci, 400hp   12598723: 427ci, 505hp
12560799: 364ci, 400hp (uu)

Head: 12564243: 364ci, 400hp   12578452: 427ci, 505hp

Abbreviations: at=automatic transmission, ci=cubic inch, hp=horsepower, mt=manual transmission, uu=possible but unlikely usage.

2006 Facts

- A 6-speed paddle shift automatic was optional. Electronically controlled by an internally-mounted, 32-bit processor, it permitted driver selection of two full automatic shifting modes, Drive and Sport, plus manual control in Sport mode by steering-wheel-mounted paddles.

- Front airbags employed passenger-sensing technology. A sensor in the passenger seat automatically disabled the passenger-side airbag if no passenger (or a passenger under a predetermined weight) was detected.

- XM satelite radio, a $325 2005 option, was included with all but base (US8) sound systems. XM's antenna was in the exterior rearview mirrors.

- A 3-spoke steering wheel replaced 2005's 4-spoke unit. The 3-spoke wheel was 16 mm (0.62-inch) smaller measured at the outside diameter.

- Z06 returned. Based on the coupe body, Z06 had unique aluminum-frame architecture, magnesium-supported fixed roof panel, magnesium engine cradle, specific body panels and trim, and a 427 cubic-inch (7.0L), 470 lb-ft, V8 mated exclusively to a 6-speed manual transmission.

- The Z06 engine (LS7), had the same 427 cubic-inch displacement as legendary "big block" 1966-1969 Corvettes, but 2006's was small-block based. Hand assembled at GM's Performance Build Center in Wixom, Michigan, the engine featured 4.125-inch cylinder bores in an aluminum block with pressed-in steel cylinder liners, titanium connecting rods and intake valves, an 8-quart dry-sump oiling system with remote reservoir, 11:1 compression, 7000-rpm redline, and 3.7-second 0-60 mph times.

- Z06 included extended mobility (run-flat) Goodyear Eagle F1 supercar tires, P275/35ZR18 front and P325/30ZR19 rear. Z06 wheels, a specific ten-spoke design, were the same diameter as coupe and convertible, but wider at 18x9.5-inch front, 19x12-inch rear. Brakes were 6-piston, 6-pad front and 4-piston, 4-pad rear with cross drilled rotors measuring 14x1.3" front and 13.4x1" rear. This compared to 12.8x1.26" front and 12x1" rear for base coupe and convertible; 13.4x1.26" front and 13x1" rear for Z51.

- Z06 had a unique front facia with larger grille and cold-air scoop. Fenders were wider and restyled. Front fenders and wheelhouses were carbon fiber. Carbon fiber sandwiched the balsa floor. The battery moved from under the hood to aft of the rear wheels. No passenger power seat adjuster was available, sound deadening was reduced and a hatch power pulldown was not included. Z06 curb weight was 3,132 lbs.

- Titanium Gray interior color replaced Steel Gray. For exteriors, Velocity Yellow replaced Millennium Yellow; Monterey Red replaced Magnetic Red. The new 2006 exterior colors were available for late 2005 production.

- A red, white and blue Z06 paced the May 28th, 2006 Indy 500.

- Instrument panel trim colors were revised. Chrome was added to HVAC and non-navigation radio knobs. A new shift knob for automatics added aluminum trim. A small GM badge was added to each rear fender. Shift-to-reverse before shutdown with manual transmission was eliminated.

2006 Options

RPO#	DESCRIPTION	QTY	RETAIL $
1YY07	Base Corvette Coupe	16,598	$44,600.00
1YY67	Base Corvette Convertible	11,151	52,335.00
1YY87	Z06 Corvette	6,272	65,800.00
2LT	Equipment Group-Coupe	1,904	1,495.00
2LZ	Preferred Equipment Group-Z06	4,854	2,900.00
3LT	Preferred Equipment Group-Coupe	10,953	4,795.00
3LT	Preferred Equipment Group-Convertible	9,972	3,395.00
C2L	Dual Removable Roof Panels (coupe)	3,726	1,400.00
CC3	Removable Roof Panel, transparent (cpe)	3,561	750.00
CM7	Power Convertible Top (3LT req'd)	8,537	1,995.00
F55	Magnetic Selective Ride Control (cpe/conv)	5,709	1,695.00
MX0	Six-speed Paddle Auto Trans (cpe/conv)	19,094	1,250.00
Q44	Competition Gray Alum Wheels (Z06)	279	295.00
QG7	Polished Aluminum Wheels (cpe/conv)	16,133	1,295.00
QL9	Polished Aluminum Wheels (Z06)	3,449	1,295.00
QX1	Comp Gray Alum Wheels (cpe/conv)	887	295.00
QX3	Chrome Aluminum Wheels (cpe/conv)	2,803	1,995.00
R8C	Corvette Museum Delivery	1,172	490.00
UE1	OnStar System (cpe/conv, 3LT req'd)	12,869	695.00
U3U	AM/FM CD DVD Nav XM ($3,340 w/1LZ)	17,474	1,600.00
US9	AM/FM 6-CD, Bose, XM (w/1LZ, nc 3LT)	10,690	1,740.00
Z51	Performance Package (cpe/conv)	10,338	1,695.00
19U	LeMans Blue exterior paint	3,459	300.00
45U	Velocity Yellow exterior paint	4,122	750.00
80U	Monterey Red exterior paint	5,052	750.00

• All models included 6-speed manual trans, stereo CD with MP3 , leather seats, 6-way power driver seat, active handling, traction control, dual-zone electronic air conditioning, keyless access and start, foglamps, tilt steering wheel, cruise control, tire pressure monitor, and floor mats.

• Coupe and convertible included 400hp LS2 engine. Z06 included 505hp LS7 engine and head-up display. Coupe and Z06 included frontal air bags. Coupe included body-color removable roof panel and power hatch release with power pulldown. Convertible included power trunk release, side airbags, sport seats, power passenger seat, cargo net and manual soft top.

• Base coupe =1LT. Base convertible = 2LT. Base Z06 = 1LZ.

• 2LT coupe included seat-mounted side impact air bags, sport seats, 6-way power passenger seat, cargo net, and luggage shade.

• 2LZ included seat-mounted side impact air bags, power-telescopic steering, auto dimming mirrors, heated seats, US9, cargo net, luggage shade, memory package, compass, and homelink garage door transmitter.

• 3LT coupe and convertible included uplevel equipment plus head-up display, power-telescopic steering column, auto-dimming inside and outside mirrors, memory package, heated seats, US9 radio, compass and homelink (garage door) transmitter.

• Magnetic Selective Ride Control (F55) was available with coupe or convertible; not available with Z51 or Z06.

• Performance Package (Z51) included specific springs, shocks and stabilizer bars, larger cross-drilled brake rotors, engine oil, transmission and power steering fluid coolers, Goodyear Eagle F1 Supercar EMT tires, and specific 6-speed manual. Also available with automatic, but unlike 2005, Z51 did not include a performance axle ratio with automatic.

2006 Colors

Code	Exterior	Coupe	Conv	Z06	Total	Interior
10	Arctic White	941	647	na	1,588	R-E-C-TG
19	LeMans Blue ✓	1,815	973	671	3,459	E-C-TG
41	Black ✓	3,365	2,206	1,672	7,243	R-E-C-TG
45	Velocity Yellow ✓	1,540	1,118	1,464	4,122	E-C-TG
67	Machine Silver ✓	2,370	1,636	978	4,984	R-E-C-TG
71	Sunset Orange ✓	930	517	282	1,729	E-T-C
74	Victory Red ✓	2,758	1,881	1,205	5,844	R-E-C-TG
80	Monterey Red	2,879	2,173	na	5,052	R-E-C-TG

✓ Available with Z06 models.

• Required combinations shown. Base seats (1LT) were Ebony only. Z06 seats were Ebony, Ebony with Titanium Gray accents, or Ebony with Red accents at no added cost.

Interior Codes: 023=Red, 193=Ebony, 313=Cashmere, 843=Titanium Gray.

Abbreviations: R=Red, E=Ebony, C=Cashmere, TG= Titanium Gray.

2007 Corvette

21,484 coupe, 10,918 conv, 8,159 Z06, 40,561 total.

2007 Numbers

Vehicle: 1G1YY25E575100001 thru 1G1YY26U675140559
  • Sixth digit is a 2 for coupe and Z06, 3 for convertible.
  • Seventh digit is restraint code: 5=front, 6=front/side
  • Eighth digit is engine code: U=364ci, 400hp (LS2), Y=427ci, 505hp (LS7) thru VIN 100161, E=427ci, 505hp (LS7) starting VIN 100162
  • Ninth digit is a security code and varies.
  • Ending VIN did not match total quantity because 2007 production included non-salable prototype builds.

Suffix: ZLF: 364ci, 400hp, at   ZLH: 364ci, 400hp, mt
       ZLD: 427ci, 505hp, mt

Block: 12568952: 364ci, 400hp   12598723: 427ci, 505hp

Head: 12564243: 364ci, 400hp   12578452: 427ci, 505hp

Abbreviations: at=automatic transmission, ci=cubic inch, hp=horsepower, mt=manual transmission.

2007 Facts

• Larger cross-drilled brake rotors were included with Magnetic Selective Ride Control (RPO F55). For previous C6s, this brake package, sized between the base and Z06 brake systems, was included and only available with the RPO Z51 Performance Package. Combined with F55, this meant larger brakes with standard tire life and weather capability.

• Air bag technology introduced in the 2006 Corvette eliminated the passenger-side on-off switch located in the glovebox. For 2007, the switch enclosure was removed, permitting a small increase in glovebox volume.

• The power convertible top, previously optional at $1,995 with the 3LT convertible equipment group, was included with 3LT for 2007. This equipment group's price increased from $3,395 to $5,540.

• OnStar, available on C6 coupes and convertibles with 3LT, was optional with Z06, but only when ordered with the 2LZ equipment group.

• Daytona Sunset Orange Metallic exterior paint was replaced by Atomic Orange Metallic. Unlike the color it replaced, Atomic Orange was extra cost due to an additional tint coat. This brought Corvette's extra cost color palette to four; LeMans Blue ($300), and Velocity Yellow, Monterey Red and Atomic Orange at $750 each.

• Previous 2006 coupe and convertible seats were monotone only. For 2007, monotone seats were standard, but two-tone seats were optional with coupe and convertible models in Ebony with Red, Ebony with Cashmere, and Ebony with Gray. The option cost was $695 and these two-tone seats included Corvette crossed flag embroidery in the headrest area and contrasting stitching.

• As in 2006, Z06 seat color choices were Ebony, Ebony with Red, or Ebony with Gray. All were available at no additional cost. For 2007, Z06 seat embroidery design was revised from outline to solid.

• Previously, Chevrolet recommended combinations of exterior, interior and convertible top colors. Dealers could often override the recommended combinations, but not in 2005 or 2006. For 2007, overrides were permitted with an extra cost option, RPO D30 at $590.

• Steering wheel controls for volume, seek, and station presets were included in the wheel's right-hand spoke for all Bose audio systems. Bose audio system enhancements enabled a +3dB maximum volume increase.

• Two special editions were introduced February 1, 2007, at the Chicago Auto Show. The Ron Fellows ALMS (American LeMans Series) GT1 Z06 honored the famed racer and was the first signed limited edition in Corvette history. It was Arctic White with Monterey Red front fender stripes and had unique interior and exterior trim. The 399 build included 300 for the US, 33 for Canada and 66 for other markets. Production of the Indy Pace car replica, an Atomic Orange convertible with Ebony seats, was limited to 500. It previewed 2008-style wheels in Sterling Silver, and included unique graphics, trim and seat upholstery.

2007 Options

RPO#	DESCRIPTION	QTY	RETAIL $
1YY07	Base Corvette Coupe	21,484	$44,995.00
1YY67	Base Corvette Convertible	10,418	52,910.00
1YY67	Indy 500 Pace Convertible (Z4Z)	500	66,995.00
1YY87	Z06 Corvette (increase to $70,000 7-26-06)	7,760	66,465.00
1YY87	Ron Fellows Z06 Special Edition (Z33)	399	77,500.00
2LT	Equipment Group-Coupe	2,606	1,495.00
2LZ	Equipment Group-Z06	6,487	3,485.00
3LT	Equipment Group-Coupe	11,934	4,945.00
3LT	Equipment Group-Convertible	9,533	5,540.00
C2L	Dual Removable Roof Panels (coupe)	3,558	1,400.00
CC3	Removable Roof Panel, transparent (coupe)	4,370	750.00
D30	Non-recommended color/trim/top combo	126	590.00
F55	Magnetic Selective Ride Control (cpe/conv)	5,619	1,995.00
MX0	Six-speed Paddle Auto Trans (cpe/conv)	22,422	1,250.00
Q44	Competition Gray Alum Wheels (Z06)	545	395.00
QG7	Polished Aluminum Wheels (cpe/conv)	3,461	1,295.00
QL9	Polished Aluminum Wheels (Z06)	3,459	1,495.00
QX1	Competition Gray Alum Wheels (cpe/conv)	1,091	395.00
QX3	Chrome Aluminum Wheels (cpe/conv)	19,850	1,850.00
R8C	Corvette Museum Delivery	1,104	490.00
UE1	OnStar System (3LT or 2LZ req'd)	18,074	695.00
U3U	AM/FM, CD, Nav, XM, Bose ($3,640 1LZ)	20,653	1,750.00
US9	AM/FM 6-CD, XM, Bose (w/1LZ))	9,970	1,890.00
Z51	Performance Package (coupe/conv)	13,696	1,695.00
19U	LeMans Blue exterior paint	3,854	300.00
45U	Velocity Yellow exterior paint	3,755	750.00
80U	Monterey Red exterior paint	5,023	750.00
83U	Atomic Orange exterior paint	3,790	750.00
**6	Two-Tone Seats with Embroidery (coupe/conv)		695.00

• All models included 6-speed manual trans, stereo CD with MP3 , leather seats, 6-way power driver seat, active handling, traction control, dual-zone air conditioning, keyless access and start, foglamps, manual tilt steering, cruise control, tire pressure monitor, and floor mats.

• Coupe and convertible included 400hp LS2 engine. Z06 included 505hp LS7 engine and head-up display. Coupe and Z06 included frontal air bags. Coupes included body-color removable roof panel and power hatch release with power pulldown. Convertibles included power trunk release, side airbags, sport seats, power passenger seat, cargo net and manual soft top.

• Base coupe=1LT, Base convertible=2LT, Base Z06=1LZ.

• 2LT coupe included seat-mounted side impact air bags, sport seats, 6-way power passenger seat, cargo net, and luggage shade.

• 2LZ included seat-mounted side impact air bags, power-telescopic steering, auto dimming mirrors, US9, cargo net, luggage shade, memory package, compass and universal home remote transmitter.

• 3LT coupe and convertible included uplevel equipment plus head-up display, power-telescopic steering column, auto-dimming mirrors, memory package, heated seats, US9 radio, compass and universal home remote transmitter. Convertibles also included power top.

• Magnetic Selective Ride Control (F55) included larger cross-drilled brake rotors; available coupe or conv; not available with Z51 or Z06.

• RPO Z51 included HD springs/shocks/stabilizers, larger cross-drilled rotors, engine oil, trans and steering coolers, Goodyear F1 Supercar EMT tires, and specific 6-speed manual. Also available with auto transmission.

2007 Colors   Quantity

Code	Exterior	Coupe	Conv	Z06	Total	Interior
10	Arctic White ✓	1,312	676	399	2,387	R-E-EC-EG-ER-C-G
19	LeMans Blue ✓	2,243	784	827	3,854	E-EC-EG-ER-C-G
41	Black ✓	4,908	2,091	2,527	9,526	R-E-EC-EG-ER-C-G
45	Velocity Yellow ✓	1,672	843	1,240	3,755	E-EC-EG-C-G
67	Machine Silver ✓	2,887	1,589	1,042	5,518	R-E-EC-EG-ER-C-G
74	Victory Red ✓	3,556	1,701	1,451	6,708	R-E-EC-EG-ER-C-G
80	Monterey Red	3,157	1,866	na	5,023	R-E-EC-EG-ER-C-G
83	Atomic Orange ✓	1,749	1,368	673	3,790	E-EG-C-G

✓Available with Z06 models (Arctic White was Fellows Z06 only).

• Recommended combos shown; override available with RPO D30. Z06 seats were Ebony, Ebony with Gray accents, or Ebony with Red accents.

Interior Codes: 023=Red, 193=Ebony, 313=Cashmere, 843=Gray, 026=Ebony/Red, 316=Ebony/Cashmere, 846=Ebony/Gray.

Abbreviations: B=Black, Be=Beige, C=Cashmere, G=Gray, E=Ebony, EC=Ebony/Cashmere, EG=Ebony/Gray, ER=Ebony/Red, R=Red.

2008 Corvette
20,030 coupe, 7,549 conv, 7,731 Z06, 35,310 total.

2008 Numbers

Vehicle: 1G1YY26W485100001 thru 1G1YY36W085135284
- Sixth digit is a 2 for coupe and Z06, 3 for convertible.
- Seventh digit is restraint code: 5=front, 6=front/side
- Eighth digit is engine code: W=376ci, 430hp (LS3); E=427ci, 505hp (LS7)
- Ninth digit is a security code and varies.
- Ending VIN did not match total quantity because 2008 production included pilot and prototype builds.

Suffix: ZAF: 376ci, 430hp, at   ZAH: 376ci, 430hp, mt
        ZAD: 427ci, 505hp, mt

Block: 12584724: 376ci, 430hp   12598723: 427ci, 505hp

Head: 12600821: 376ci, 430hp   12600823: 376ci, 430hp, uu
       12578452: 427ci, 505hp

Abbreviations: at=automatic transmission, ci=cubic inch, hp=horsepower, mt=manual transmission, uu=possible but unlikely usage.

2008 Facts

- The base engine for 2008 was the new LS3. It increased from 364 (6.0 L) to 376-ci (6.2L), and from 400 to 430 horsepower. Cylinder heads had Z06-style large-valve, large-port design with larger, straighter intake ports. Intake valves had hollow stems with nine-percent larger diameters. The camshaft had five-percent greater lift.

- LS3's beauty covers and composite intake manifolds had acoustically tuned sections to dampen and tune valve train noise.

- An optional Dual Mode Exhaust System (RPO NPP) for coupes and convertibles used vacuum-actuated outlet valves controlled by rpm and throttle position. Pipe diameter was 2.5-inches; exhaust note was more aggressive. LS3's output increased to 436-horsepower with NPP.

- Steering feel was improved by more precise component machining, a stiffer intermediate shaft, and revised controller calibrations.

- Manual transmission linkage had shorter fore-aft travel. The automatic had hardware and software changes for quicker manual paddle shifts. Also, a 2.73:1 performance axle ratio was optional with automatics, and included with the combination of automatic and Z51.

- A new split-spoke wheel was included with coupes and convertibles. The base wheel was painted Sparkle Silver (it debuted with the 2007 Indy Pace replica with a Sterling Silver paint finish); Competition Gray paint was optional. A five-spoke forged wheel was optional, polished or chromed. The 2005-07 wheel style was optional in chrome. Z06 wheels weren't changed except for the 427-Limited Edition Z06.

- A new remote fob concealed its key in a sliding compartment.

- Interiors were revised with a new wrapped "cyber" graphic pattern console trim plate, bright surrounds for the cupholder and manual shifter, and brushed aluminum shifter and door release trim. Metal door sill plates were new (interim 2007). Floor mat retainers were simplified.

- 3LZ&4LT combined uplevel group features with leather-wrapped instrument panel, door panels, and console. Console trim had unique bias-graphic pattern. 4LT had crossed-flag headrest embroidery. Available with all models in Linen, Sienna and Ebony.

- OnStar, XM radio, and auto-dimming rearview mirror with compass were standard. All radios except navigation had an MP3 audio jack.

- The 2008 Indy 500 was paced by a Z06 burning E85 fuel. A second pace car design was available in silver/black to honor the famous 1978 model. Available in coupe or convertible, all had Titanium seats and Z06 spoilers. Each was numbered and signed by Emerson Fittipaldi.

- A 427-Limited Edition Z06 had Crystal Red paint, Dark Titanium custom interior, "spider" chrome wheels and special trim/graphics. All 505 were signed by retired Bowling Green Corvette Plant Manager Wil Cooksey.

- 500 RPO ZHZ coupes were built for Hertz Fun Club rental. All had Velocity yellow paint with black stripes and seven-spoke chrome wheels.

2008 Options

RPO#	DESCRIPTION	QTY	RETAIL $
1YY07	Base Corvette Coupe	19,796	$45,995.00
1YY67	Base Corvette Convertible	7,283	54,335.00
1YY07	Indy 500 Pace Coupe (Z4Z)	234	59,090.00
1YY67	Indy 500 Pace Convertible (Z4Z)	266	68,160.00
1YY87	Z06 Corvette	7,226	71,000.00
1YY87	427 Limited Edition Z06 (Z44)	505	84,195.00
2LT	Equipment Group-Coupe	3,086	1,495.00
2LZ	Equipment Group-Z06	4,929	3,045.00
3LT	Equipment Group-Coupe	9,201	4,505.00
3LT	Equipment Group-Convertible	5,879	5,100.00
3LZ	Equipment Group-Z06	1,460	6,545.00
4LT	Equipment Group-Convertible	528	8,600.00
4LT	Equipment Group-Coupe	834	8,005.00
C2L	Dual Removable Roof Panels (coupe)	2,773	1,400.00
CC3	Removable Roof Panel, transparent (cpe)	3,251	750.00
D30	Non-recommended color/trim/top combo	189	590.00
F55	Magnetic Selective Ride Control (cpe/conv)	4,666	1,995.00
GU2	Rear Axle 2.73 Ratio (w/auto trans)	8,839	395.00
MX0	Six-speed Paddle Auto Trans (cpe/conv)	19,136	1,250.00
NPP	Dual Mode Exhaust System (cpe/conv)	13,454	1,195.00
Q44	Competition Gray Aluminum Wheels (Z06) ...	395	395.00
Q76	Chrome Aluminum Wheels (Z06)	5,101	1,995.00
Q9V	Chrome Forged Alum Wheels (cpe/conv)	2,932	1,850.00
QG7	Polished Forged Alum Wheels (cpe/conv)	5,412	1,295.00
QL9	Polished Aluminum Wheels (Z06)	970	1,495.00
QX1	Competition Gray Alum Wheels (cpe/conv) ...	1,781	395.00
QX3	Original Design Chrome Wheels (cpe/conv) ...	9,626	1,850.00
R8C	Corvette Museum Delivery	954	490.00
U3U	AM/FM, CD, Navigation, Bose	16,807	1,750.00
Z51	Performance Package (coupe/conv)	10,706	1,695.00
45U	Velocity Yellow exterior paint	3,264	750.00
83U	Atomic Orange exterior paint	2,246	300.00
85U	Jetstream Blue exterior paint	3,728	750.00
89U	Crystal Red exterior paint	5,420	750.00
**6	Modified Two-Tone Seats (2LT&3LT)	5,807	695.00

• All included 6-speed manual, stereo CD, XM, MP3 jack, OnStar, leather seats, power driver seat, active handling, traction control, dual-zone a/c, keyless access/start, foglamps, tilt, cruise control, tire pressure monitor.

• Coupe and convertible included 430hp LS3 engine. Z06 included 505hp LS7 engine and head-up display. Coupe and Z06 included frontal air bags. Coupes included body-color removable roof panel and power hatch release with power pulldown. Convertibles included power trunk release, side airbags, sport seats, power passenger seat, cargo net, and manual soft top.

• 2LT coupe included seat-mounted side impact air bags, sport seats, 6-way power passenger seat, cargo net, and luggage shade.

• 2LZ had seat-mounted side impact air bags, power-telescopic steering, heated seats, cargo net, luggage shade, memory package, steering-wheel audio controls, US9 6-CD radio, enhanced acoustics, and universal home remote transmitter.

• 3LT coupe and convertible included uplevel equipment plus head-up display, power-telescopic steering column, memory package, heated seats, US9 6-CD radio, steering-wheel audio controls, and universal home remote transmitter. Convertibles also included power top.

• 3LZ/4LT had uplevel equipment plus leather-wrapped interior and bias-pattern console trim. 4LT had crossed-flag embroidery on headrests.

2008 Colors   Quantity

Code	Exterior	Coupe	Conv	Z06	Total	Interior
10	Arctic White	1,320	654	na	1,974	R-E-EC-EG-ER-C-TG-S-L
41	Black	4,977	1,818	2,585	9,380	R-E-EC-EG-ER-C-TG-S-L
45	Velocity Yellow	1,816	550	898	3,264	E-EC-EG-C-TG-L
67	Machine Silver	2,088	753	819	3,660	R-E-EC-EG-ER-C-TG-S-L
74	Victory Red	3,120	1,116	1,402	5,638	R-E-EC-EG-ER-C-TG-L
83	Atomic Orange	1,272	458	516	2,246	E-EG-C-TG-S-L
85	Jetstream Blue	2,067	655	1,006	3,728	E-EC-EG-ER-C-TG-S-L
89	Crystal Red	3,370	1,545	505	5,420	R-E-EC-EG-ER-C-TG-S-L-DT

• Recommended combos shown; override available with RPO D30.

Interior Codes: 023=Red, 026=Ebony/Red, 193=Ebony, 195=Ebony custom, 245=Sienna custom, 313=Cashmere, 316=Ebony/Cashmere, 425=Linen custom, 785=Dark Titanium custom, 843=Titanium Gray, 846=Ebony/Gray.

Abbreviations: B=Black, Be=Beige, C=Cashmere, DT=Dark Titanium, E=Ebony, EC=Ebony/Cashmere, EG=Ebony/Gray, ER=Ebony/Red, G=Gray, L=Linen, R=Red, S=Sienna, TG=Titanium Gray,

2009 Corvette

2009 Numbers

Vehicle: 1G1YY26W295100001 thru 1G1YY------------
1G1YR26R695800001 thru 1G1YY------------ (ZR1)
1G1YY25R695700001 thru 1G1YY26R695700060 (ZR1 pilot)
1G1YR26R495900001 (Special ZR1 auction build)
- Sixth digit is a 2 for coupe, Z06 and ZR1, 3 for convertible.
- Seventh digit is restraint code: 5=front, 6=front/side
- Eighth digit is engine code: W=376ci, 430hp (LS3); E=427ci, 505hp (LS7), R=376ci, 638hp (LS9).
- Ninth digit is a security code and varies.

Suffix: ZAJ: 376ci, 430hp, at   ZAK: 376ci, 430hp, mt
        SAY: 427ci, 505hp, mt   ZAM: 376ci, 638hp, mt

Block: 12621766: 376ci, 430hp and 638hp   12598723: 427ci, 505hp

Head: 12600821: 376ci, 430hp   12600823: 376ci, 430hp, uu
       12578452: 427ci, 505hp   12621771: 376ci, 638hp

Abbreviations: at=automatic transmission, ci=cubic inch, hp=horsepower, mt=manual transmission, uu=possible but unlikely usage.

2009 Facts

- Corvette's 200+-mph supercar made its debut as the 2009 ZR1. With 638 horsepower, it was the most powerful automobile ever sold by GM.

- ZR1's LS9 engine had the same 376 cubic-inch (6.2L) displacement as Corvette's base LS3, but the LS9 was a new design with a 10.5-quart dry sump oil lubrication system. A positive-displacement Roots-type supercharger with a four-lobe rotor design provided boost. Compression ratio was 9.1:1. Premium fuel was required.

- ZR1s utilized a twin-disc clutch for increased clamping power. A stronger 6-speed manual was included. An automatic was not available.

- ZR1's chassis with aluminum frame was Z06-derived, but ZR1 had Selective Magnetic Ride Control, specially tuned for this application.

- ZR1's body was also derived from Z06's fixed roof coupe architecture, but with notable differences. ZR1's raised hood was carbon-fiber with a clear, polycarbonate center window to show the top of the supercharger's intercooler. The carbon fiber hood was painted body-color with the carbon fiber weave evident on its underside. Front fenders were carbon fiber, painted body color, with unique side ports. Other carbon fiber parts, the fixed roof panel, roof bow, front facia splitter and rocker moldings, were unpainted carbon fiber with a protective clear coating.

- ZR1s had massive Brembo brakes. Rotors were vented and drilled carbon-fiber-reinforced ceramic silicon carbide. Front rotors were 15.5-inches, rear 15-inches. Front calipers were six-piston, rears were four.

- Another ZR1 Corvette first was the standard Michelin tire package. Pilot Sport 2 front tires were P285/30ZR19; rear were P335/25ZR20.

- Steering wheel audio controls were standard with all 2009 models.

- Cyber Gray and Blade Silver were new exterior colors. Custom interior choices expanded to include Ebony (late 2008) and Dark Titanium which was first used with the 2008 427-Limited Edition Z06.

- All models included variable-ratio power steering. Steel steering column shafts replaced aluminum for more precise control.

- The Z06's LS7 engine remained at 505hp, but the oil capacity of its dry-sump system was increased from 8 to 10.5-quarts, matching ZR1's LS9. Also, 2009 Z06s (and ZR1s) had the power hatch pulldown feature which previously had been exclusive to C6 base coupes.

- Ten wheel designs and/or wheel finishes were factory-available for Corvette's four models, the most ever. The coupe and convertible base wheel was a silver-painted five split-spoke design; a gray-painted version optional. A forged five spoke design was optional in polished or chrome finish. Z06 wheels were a new spider design, silver-painted standard with gray-painted and chrome versions optional. The previous Z06 five wide-split spoke wheel was still optional in chrome. ZR1's came with an exclusive 20-spoke design painted silver, also optional with chrome finish.

2009 Options

RPO#	DESCRIPTION	RETAIL $
1YY07	Base Corvette Coupe	$47,895.00
1YY67	Base Corvette Convertible	52,550.00
1YY87	Z06 Corvette	73,255.00
1YY87	ZR1 Corvette	103,300.00
2LT	Equipment Group-Coupe	1,545.00
2LT	Equipment Group-Convertible	3,540.00
2LZ	Equipment Group-Z06	3,015.00
3LT	Equipment Group-Coupe	4,555.00
3LT	Equipment Group-Convertible	6,550.00
3LZ	Equipment Group-Z06	6,515.00
3ZR	Equipment Group-ZR1	10,000.00
4LT	Equipment Group-Convertible	10,050.00
4LT	Equipment Group-Coupe	8,055.00
C2L	Dual Removable Roof Panels (coupe)	1,400.00
CC3	Removable Roof Panel, transparent (coupe)	750.00
D30	Non-recommended color/trim/top combo	590.00
F55	Magnetic Selective Ride Control (coupe/conv)	1,995.00
GU2	Rear Axle 2.73 Ratio (w/auto trans)	395.00
MX0	Six-speed Paddle Auto Trans (coupe/conv)	1,250.00
NPP	Dual Mode Exhaust System (coupe/conv)	1,195.00
Q44	Competition Gray Aluminum Wheels (Z06)	395.00
Q6B	Chrome 20-spoke Aluminum Wheels (ZR1)	2,000.00
Q76	Chrome Aluminum Wheels (Z06)	1,995.00
Q8A	Spider Chrome Aluminum Wheels (Z06)	1,995.00
Q9V	Forged Chrome Aluminum Wheels (cpe/conv)	1,850.00
QG7	Forged Polished Aluminum Wheels (cpe/conv)	1,295.00
QL9	Polished Aluminum Wheels (Z06)	1,495.00
QX1	Competition Gray Aluminum Wheels (cpe/conv)	395.00
R8C	Corvette Museum Delivery	490.00
R8E	Gas Guzzler Tax (ZR1)	1,700.00
U3U	AM/FM, CD, Navigation, Bose	1,750.00
VPK	Exterior Appearance Package	2,095.00
VPL	Exterior Appearance Package w/wheels	5,475.00
Z51	Performance Package (coupe/conv)	1,695.00
45U	Velocity Yellow exterior paint	750.00
83U	Atomic Orange exterior paint	300.00
85U	Jetstream Blue exterior paint	750.00
89U	Crystal Red exterior paint	750.00
**6	Modified Two-Tone Seats w/embroidery	695.00

• All included 6-speed manual stereo CD, XM, MP3 jack, OnStar, leather seats, power driver seat, active handling, traction control, dual-zone a/c, keyless access/start, foglamps, tilt, cruise control, tire pressure monitor.
• Coupe and convertible included 430hp LS3 engine. Z06 included 505hp LS7 engine. ZR1 included 638hp LS9 engine and magnetic ride control. Coupes included body-color removable roof panel. Coupe, Z06, and ZR1 included power hatch release with power pulldown. Convertibles included power trunk release and manual soft top.
• Many features were available only in equipment groups including side air bags, memory package, Bluetooth phone connection (new for 2009), power passenger seat, custom leather interior, heated seats, power telescopic steering column, universal home remote and power top for convertibles.

2009 Colors

Code	Exterior	Soft Top	Suggested Interiors
10	Arctic White	B-Be-G	R-C-E-EC-EG-ER-DT-L-S-TG
17	Blade Silver ✓	B-Be-G	R-C-E-EC-EG-ER-DT-L-S-TG
41	Black ✓	B-Be-G	R-C-E-EC-EG-ER-DT-L-S-TG
45	Velocity Yellow ✓	B-Be	R-C-E-EC-EG-ER-DT-L-S-TG
57	Cyber Gray ✓	B-Be-G	R-C-E-EC-EG-ER-DT-L-S-TG
74	Victory Red ✓	B-Be-G	R-C-E-EC-EG-ER-DT-L-S-TG
83	Atomic Orange ✓	B-Be	R-C-E-EC-EG-ER-DT-L-S-TG
85	Jetstream Blue ✓	B-Be-G	R-C-E-EC-EG-ER-DT-L-S-TG
89	Crystal Red	B-Be	R-C-E-EC-EG-ER-DT-L-S-TG

• Recommended combos shown; override available with RPO D30.
✓ Available with Z06 and ZR1 models.

Interior Codes: 023=Red, 026=Ebony/Red,193=Ebony, 195=Ebony custom, 245=Sienna custom, 313=Cashmere, 316=Ebony/Cashmere, 425=Linen custom, 785=Dark Titanium custom, 843=Titanium Gray, 846=Ebony/Gray.

Abbreviations: B=Black, Be=Beige, C=Cashmere, DT=Dark Titanium, E=Ebony, EC=Ebony/Cashmere, EG=Ebony/Gray, ER=Ebony/Red, G=Gray, L=Linen, R=Red, S=Sienna, TG=Titanium Gray.

Notes

Notes

1953 Corvette roadster

Author photos

Length: 167.3" overall
Width: 69.8 "
Height: 51.5"over windshield
Curb wgt: 2,886 lbs.
Wheelbase: 102"
Tire size: 6.70x15"
Front track: 57"
Rear track: 58.8"
Fuel tank: 18.0 gal. (approx)

1954 Corvette roadster

Length: 167.3" overall
Width: 69.8 "
Height: 51.5"over windshield
Curb wgt: 2,886 lbs.
Wheelbase: 102"
Tire size: 6.70x15"
Front track: 57"
Rear track: 58.8"
Fuel tank: 17.25 gal.

1955 Corvette roadster

Author photos

Length: 167.3" overall
Width: 69.8 "
Height: 51.5"over windshield
Curb wgt: 2,805 lbs.
Wheelbase: 102"
Tire size: 6.70x15"
Front track: 57"
Rear track: 58.8"
Fuel tank: 17.25 gal.

1956 Corvette convertible

Author photos

Length: 168″ overall
Width: 70.5″
Height: 51.5″ over hardtop
Curb wgt: 2,886 lbs.
Wheelbase: 102″
Tire size: 6.70x15″
Front track: 57″
Rear track: 58.8″
Fuel tank: 16.4 gal.

1957 Corvette convertible

Author photos

Length: 168″ overall
Width: 70.5″
Height: 51″ over hardtop
Curb wgt: 2,849 lbs.
Wheelbase: 102″
Tire size: 6.70x15″
Front track: 57″
Rear track: 58.8″
Fuel tank: 16.4 gal.

1958 Corvette convertible

Chevrolet photos

Length: 177.2″ overall
Width: 72.8″
Height: 51″ over hardtop
Curb wgt: 2,926 lbs.
Wheelbase: 102″
Tire size: 6.70x15″
Front track: 57″
Rear track: 58.8″
Fuel tank: 16.4 gal.

1959 Corvette convertible

Author photos

Length: 177.2" overall
Width: 72.8"
Height: 51" over hardtop
Curb wgt: 2,975 lbs.
Wheelbase: 102"
Tire size: 6.70x15"
Front track: 57"
Rear track: 58.8"
Fuel tank: 16.4 gal (24 opt)

1960 Corvette convertible

Chevrolet photos

Length: 177.2" overall
Width: 72.8"
Height: 51" over hardtop
Curb wgt: 2,985 lbs.
Wheelbase: 102"
Tire size: 6.70x15"
Front track: 57"
Rear track: 58.8"
Fuel tank: 16.4 gal (24 opt)

1961 Corvette convertible

Chevrolet photos

Length: 177.2" overall
Width: 72.8"
Height: 51.5" over hardtop
Curb wgt: 2,985 lbs.
Wheelbase: 102"
Tire size: 6.70x15"
Front track: 57"
Rear track: 58.8"
Fuel tank: 16.4 gal (24 opt)

1962 Corvette convertible

John Amgwert
photos

Length: 177.2" overall
Width: 70.4"
Height: 51.5" over hardtop
Curb wgt: 3,065 lbs.
Wheelbase: 102"
Tire size: 6.70x15"
Front track: 57
Rear track: 58.8"
Fuel tank: 16.4 gal.

1963 Corvette Split Window coupe

Author photos

Length: 175.1" overall
Width: 69.6"
Height: 49.8 (coupe)
Curb wgt: 3,015 lbs (coupe)
Wheelbase: 98"
Tire size: 6.70x15"
Front track: 56.25"
Rear track: 57"
Fuel tank: 20 gal (36 opt)

1964 Corvette coupe

Author photos

Length: 175.1" overall
Width: 69.6"
Height: 49.8 (coupe)
Curb wgt: 3,125 lbs (coupe)
Wheelbase: 98"
Tire size: 6.70x15"
Front track: 56.25"
Rear track: 57"
Fuel tank: 20 gal (36 opt)

1965 Corvette convertible

Author photos

Length: 175.1" overall
Width: 69.6"
Height: 49.8 (coupe)
Curb wgt: 3,135 lbs (coupe)
Wheelbase: 98"
Tire size: 7.75x15"
Front track: 56.8"
Rear track: 57.6"
Fuel tank: 20 gal (36 opt)

1966 Corvette convertible

Chevrolet photos

Length: 175.1" overall
Width: 69.6"
Height: 49.8 (coupe)
Curb wgt: 3,140 lbs (coupe)
Wheelbase: 98"
Tire size: 7.75x15"
Front track: 56.8"
Rear track: 57.6"
Fuel tank: 20 gal (36 opt)

1967 Corvette coupe

Author photos

Length: 175.1" overall
Width: 69.6"
Height: 49.8 (coupe)
Curb wgt: 3,155 lbs (coupe)
Wheelbase: 98"
Tire size: 7.75x15"
Front track: 56.8"
Rear track: 57.6"
Fuel tank: 20 gal (36 opt)

1968 Corvette convertible

Author photos

Length: 182.5″ overall
Width: 69″
Height: 47.8 (coupe)
Curb wgt: 3,210 lbs (coupe)
Wheelbase: 98″
Tire size: F70x15″
Front track: 58.3″
Rear track: 59″
Fuel tank: 20 gal

1969 Corvette coupe

Author photos

Length: 182.5″ overall
Width: 69″
Height: 47.8 (coupe)
Curb wgt: 3,245 lbs (coupe)
Wheelbase: 98″
Tire size: F70x15″
Front track: 58.7″
Rear track: 59.4″
Fuel tank: 20 gal

1970 Corvette coupe

Author photos

Length: 182.5″ overall
Width: 69″
Height: 47.8 (coupe)
Curb wgt: 3,285 lbs (coupe)
Wheelbase: 98″
Tire size: F70x15″
Front track: 58.7″
Rear track: 59.4″
Fuel tank: 20 gal (17 Calif)

1971 Corvette coupe

Chevrolet photos

Length: 182.5" overall
Width: 69"
Height: 47.8 (coupe)
Curb wgt: 3,202 lbs (coupe)
Wheelbase: 98"
Tire size: F70x15"
Front track: 58.7"
Rear track: 59.4"
Fuel tank: 18 gal

1972 Corvette coupe

Chevrolet photos

Length: 182.5" overall
Width: 69"
Height: 47.8 (coupe)
Curb wgt: 3,305 lbs (coupe)
Wheelbase: 98"
Tire size: F70x15"
Front track: 58.7"
Rear track: 59.4"
Fuel tank: 18 gal

1973 Corvette coupe

Author photos

Length: 184.6" overall
Width: 69"
Height: 47.8 (coupe)
Curb wgt: 3,416 lbs (coupe)
Wheelbase: 98"
Tire size: GR70x15"
Front track: 58.7"
Rear track: 59.5"
Fuel tank: 18 gal

1974 Corvette coupe

Author photos

Length: 185.5" overall
Width: 69"
Height: 47.8 (coupe)
Curb wgt: 3,388 lbs (coupe)
Wheelbase: 98"
Tire size: GR70x15"
Front track: 58.7"
Rear track: 59.5"
Fuel tank: 18 gal

1975 Corvette convertible

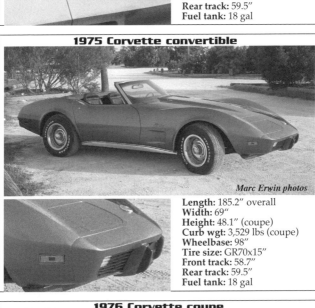

Marc Erwin photos

Length: 185.2" overall
Width: 69"
Height: 48.1" (coupe)
Curb wgt: 3,529 lbs (coupe)
Wheelbase: 98"
Tire size: GR70x15"
Front track: 58.7"
Rear track: 59.5"
Fuel tank: 18 gal

1976 Corvette coupe

Chevrolet photos

Length: 185.2" overall
Width: 69"
Height: 48.1"
Curb wgt: 3,541 lbs
Wheelbase: 98"
Tire size: GR70x15"
Front track: 58.7"
Rear track: 59.5"
Fuel tank: 17 gal

1977 Corvette coupe

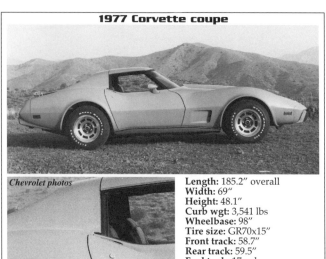

Chevrolet photos

Length: 185.2" overall
Width: 69"
Height: 48.1"
Curb wgt: 3,541 lbs
Wheelbase: 98"
Tire size: GR70x15"
Front track: 58.7"
Rear track: 59.5"
Fuel tank: 17 gal

1978 Corvette coupe (pace car)

Ed Olson photos

Length: 185.2" overall
Width: 69"
Height: 48"
Curb wgt: 3,572 lbs
Wheelbase: 98"
Tire size: P225/70Rx15"
Front track: 58.7"
Rear track: 59.5"
Fuel tank: 23.7 gal

1979 Corvette coupe

Chevrolet photos

Length: 185.2" overall
Width: 69"
Height: 48"
Curb wgt: 3,503 lbs
Wheelbase: 98"
Tire size: P225/70Rx15"
Front track: 58.7"
Rear track: 59.5"
Fuel tank: 24 gal

1980 Corvette coupe

Author photos

Length: 185.3" overall
Width: 69"
Height: 48"
Curb wgt: 3,336 lbs
Wheelbase: 98"
Tire size: P225/70Rx15"
Front track: 58.7"
Rear track: 59.5"
Fuel tank: 23.7 gal

1981 Corvette coupe

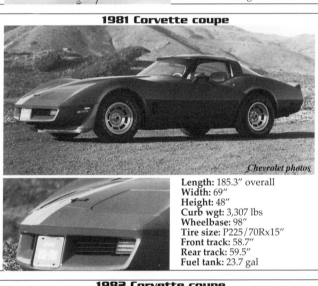

Chevrolet photos

Length: 185.3" overall
Width: 69"
Height: 48"
Curb wgt: 3,307 lbs
Wheelbase: 98"
Tire size: P225/70Rx15"
Front track: 58.7"
Rear track: 59.5"
Fuel tank: 23.7 gal

1982 Corvette coupe

Author photos

Length: 185.3" overall
Width: 69"
Height: 48"
Curb wgt: 3,342 lbs
Wheelbase: 98"
Tire size: P225/70Rx15"
Front track: 58.7"
Rear track: 59.5"
Fuel tank: 23.7 gal

1984 Corvette coupe

Author photos

Length: 176.5" overall
Width: 71"
Height: 46.7"
Curb wgt: 3,192 lbs
Wheelbase: 96.2"
Tire size: P255/50VR16"
Front track: 59.6"
Rear track: 60.4"
Fuel tank: 20 gal

1985 Corvette coupe

Author photos

Length: 176.5" overall
Width: 71"
Height: 46.7"
Curb wgt: 3,224 lbs
Wheelbase: 96.2"
Tire size: P255/50VR16"
Front track: 59.6"
Rear track: 60.4"
Fuel tank: 20 gal

1986 Corvette convertible

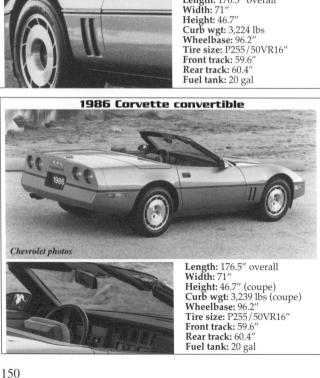

Chevrolet photos

Length: 176.5" overall
Width: 71"
Height: 46.7" (coupe)
Curb wgt: 3,239 lbs (coupe)
Wheelbase: 96.2"
Tire size: P255/50VR16"
Front track: 59.6"
Rear track: 60.4"
Fuel tank: 20 gal

1987 Corvette coupe

Author photos

Length: 176.5" overall
Width: 71"
Height: 46.7" (coupe)
Curb wgt: 3,216 lbs (coupe)
Wheelbase: 96.2"
Tire size: P255/50VR16"
Front track: 59.6"
Rear track: 60.4"
Fuel tank: 20 gal

1988 Corvette coupe

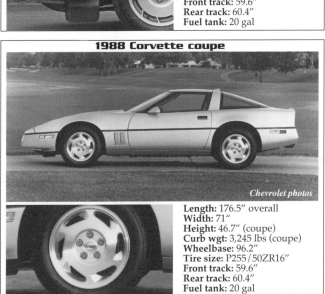

Chevrolet photos

Length: 176.5" overall
Width: 71"
Height: 46.7" (coupe)
Curb wgt: 3,245 lbs (coupe)
Wheelbase: 96.2"
Tire size: P255/50ZR16"
Front track: 59.6"
Rear track: 60.4"
Fuel tank: 20 gal

1989 Corvette coupe

Chevrolet photos

Length: 176.5" overall
Width: 71"
Height: 46.7" (coupe)
Curb wgt: 3,238 lbs (coupe)
Wheelbase: 96.2"
Tire size: P275/40ZR17"
Front track: 59.6"
Rear track: 60.4"
Fuel tank: 20 gal

1990 Corvette convertible

Chevrolet photos

Length: 176.5" overall
Width: 71"
Height: 46.7" (coupe)
Curb wgt: 3,288 lbs (coupe)
Wheelbase: 96.2"
Tire size: P275/40ZR17"
Front track: 59.6"
Rear track: 60.4"
Fuel tank: 20 gal

1991 Corvette coupe

Chevrolet photos

Length: 178.6" overall
Width: 71"
Height: 46.7" (coupe)
Curb wgt: 3,288 lbs (coupe)
Wheelbase: 96.2"
Tire size: P275/40ZR17"
Front track: 59.6"
Rear track: 60.4"
Fuel tank: 20 gal

1992 Corvette convertible

Chevrolet photos

Length: 178.5" overall
Width: 70.7"
Height: 46.3" (coupe)
Curb wgt: 3,338 lbs (coupe)
Wheelbase: 96.2"
Tire size: P275/40ZR17"
Front track: 57.7"
Rear track: 59.1"
Fuel tank: 20 gal

1993 Corvette coupe

Chevrolet photos

Length: 178.5" overall
Width: 70.7"
Height: 46.3" (coupe)
Curb wgt: 3,333 lbs (coupe)
Wheelbase: 96.2"
Tire size (F): P255/45ZR17"
Tire size (R): P285/40ZR17"
Front track: 57.7"
Rear track: 59.1"
Fuel tank: 20 gal

1994 Corvette coupe

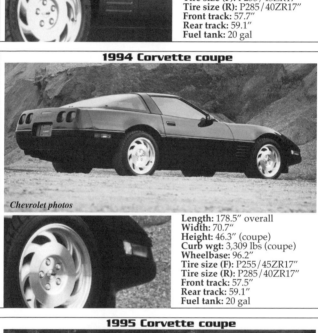

Chevrolet photos

Length: 178.5" overall
Width: 70.7"
Height: 46.3" (coupe)
Curb wgt: 3,309 lbs (coupe)
Wheelbase: 96.2"
Tire size (F): P255/45ZR17"
Tire size (R): P285/40ZR17"
Front track: 57.5"
Rear track: 59.1"
Fuel tank: 20 gal

1995 Corvette coupe

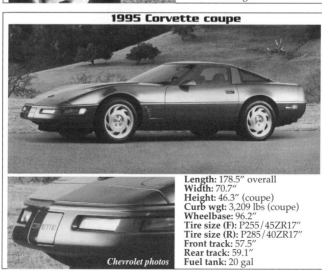

Chevrolet photos

Length: 178.5" overall
Width: 70.7"
Height: 46.3" (coupe)
Curb wgt: 3,209 lbs (coupe)
Wheelbase: 96.2"
Tire size (F): P255/45ZR17"
Tire size (R): P285/40ZR17"
Front track: 57.5"
Rear track: 59.1"
Fuel tank: 20 gal

153

1996 Corvette convertible

Chevrolet photos

Length: 178.5″ overall
Width: 70.7″
Height: 46.3″ (coupe)
Curb wgt: 3,298 lbs (coupe)
Wheelbase: 96.2″
Tire size (F): P255/45ZR17″
Tire size (R): P285/40ZR17″
Front track: 59.5″
Rear track: 59.1″
Fuel tank: 20 gal

1997 Corvette coupe

Chevrolet photos

Length: 179.6″ overall
Width: 73.6″
Height: 47.8″
Curb wgt: 3,221 lbs
Wheelbase: 104.5″
Tire size (F): P245/45ZR17″
Tire size (R): P275/40ZR18″
Front track: 62.1″
Rear track: 62″
Fuel tank: 19.1 gal

1998 Corvette convertible

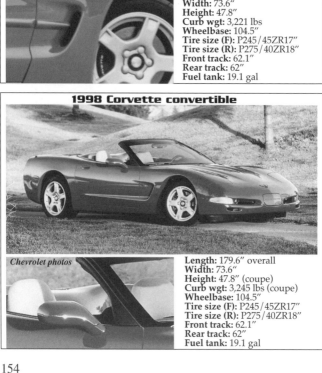

Chevrolet photos

Length: 179.6″ overall
Width: 73.6″
Height: 47.8″ (coupe)
Curb wgt: 3,245 lbs (coupe)
Wheelbase: 104.5″
Tire size (F): P245/45ZR17″
Tire size (R): P275/40ZR18″
Front track: 62.1″
Rear track: 62″
Fuel tank: 19.1 gal

1999 Corvette hardtop

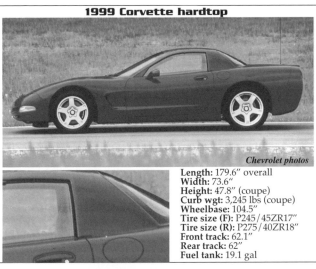

Chevrolet photos

Length: 179.6" overall
Width: 73.6"
Height: 47.8" (coupe)
Curb wgt: 3,245 lbs (coupe)
Wheelbase: 104.5"
Tire size (F): P245/45ZR17"
Tire size (R): P275/40ZR18"
Front track: 62.1"
Rear track: 62"
Fuel tank: 19.1 gal

2000 Corvette coupe

Chevrolet photos

Length: 179.6" overall
Width: 73.6"
Height: 47.8"
Curb wgt: 3,246 lbs
Wheelbase: 104.5"
Tire size (F): P245/45ZR17"
Tire size (R): P275/40ZR18"
Front track: 62.1"
Rear track: 62"
Fuel tank: 18.5 gal

2001 Corvette convertible

Chevrolet photos

Length: 179.6" overall
Width: 73.6"
Height: 47.8"
Curb wgt: 3,207 lbs
Wheelbase: 104.5"
Tire size (F): P245/45ZR17"
Tire size (R): P275/40ZR18"
Front track: 62.1"
Rear track: front, 62"
Fuel tank: 18.5 gal

2002 Corvette Z06

Chevrolet photos

Length: 179.6" overall
Width: 73.6"
Height: 47.7"
Curb wgt: 3,116 lbs
Wheelbase: 104.5"
Tire size (F): P265/40ZR17"
Tire size (R): P295/35ZR18"
Front track: 62.4"
Rear track: 62.6"
Fuel tank: 18.5 gal

2003 Corvette coupe

Chevrolet photos

Length: 179.7" overall
Width: 73.6"
Height: 47.7"
Curb wgt: 3,246 lbs
Wheelbase: 104.5"
Tire size (F): P245/45ZR17"
Tire size (R): P275/40ZR18"
Front track: 61.9"
Rear track: 62.0"
Fuel tank: 18.0 gal

2004 Corvette Z06

Chevrolet photos

Length: 179.7" overall
Width: 73.6"
Height: 47.7"
Curb wgt: 3,106 lbs (Comm)
Wheelbase: 104.5"
Tire size (F): P265/40ZR17"
Tire size (R): P295/35ZR18"
Front track: 62.4"
Rear track: 62.6"
Fuel tank: 18.0 gal

2005 Corvette coupe

Chevrolet photos

Length: 174.6" overall
Width: 72.6"
Height: 49.1"
Curb wgt: 3,179 lbs
Wheelbase: 105.7"
Tire size (F): P245/40ZR18"
Tire size (R): P285/35ZR19"
Front track: 62.1"
Rear track: 60.7"
Fuel tank: 18.0 gal

2006 Corvette Z06

Chevrolet photos

Length: 175.6" overall
Width: 75.9"
Height: 49"
Curb wgt: 3,132 lbs
Wheelbase: 105.7"
Tire size (F): P275/35ZR18"
Tire size (R): P325/30ZR19"
Front track: 63.5"
Rear track: 62.5"
Fuel tank: 18.0 gal

2007 Corvette convertible

Chevrolet photos

Length: 174.6" overall
Width: 72.6"
Height: 49"
Curb wgt: 3,199 lbs
Wheelbase: 105.7"
Tire size (F): P245/40ZR18"
Tire size (R): P285/35ZR19"
Front track: 62.1"
Rear track: 66.7"
Fuel tank: 18.0 gal

2008 Corvette coupe

Length: 174.6" overall
Width: 72.6"
Height: 49.0"
Curb wgt: 3,217 lbs
Wheelbase: 105.7"
Tire size (F): P245/40ZR18"
Tire size (R): P285/35ZR19"
Front track: 62.1"
Rear track: 60.7"
Fuel tank: 18.0 gallons

Chevrolet photos

2009 Corvette ZR1

Chevrolet photos

Length: 174.2" overall
Width: 75.9"
Height: 49.0"
Curb wgt: 3,324 lbs
Wheelbase: 105.7"
Tire size (F): P285/30ZR19"
Tire size (R): P335/25ZR20"
Front track: 63.5"
Rear track: 62.5"
Fuel tank: 18.0 gallons

BLACK BOOK ORDER FORM

Corvette Black Book 1953-2009

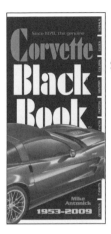

Send _____ copies @ $17.99 each $_____

Ohio residents add 1.25 sales tax_____

Postage/hard shipping container **3.00**

Check or money order enclosed $_____

Mail order to:
**Michael Bruce Associates, Inc
Post Office Box 1966
Gambier, Ohio 43022**

Name _____

Street _____

City _____State_____Zip_____

BLACK BOOK ORDER FORM

Corvette Black Book 1953-2009

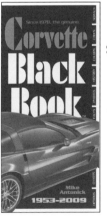

Send _____ copies @ $17.99 each $_____

Ohio residents add 1.25 sales tax _____

Postage/hard shipping container **3.00**

Check or money order enclosed $_____

Mail order to:
**Michael Bruce Associates, Inc
Post Office Box 1966
Gambier, Ohio 43022**

Name _____

Street _____

City _____ State_____ Zip_____

BLACK BOOK ORDER FORM

BLACK BOOK ORDER FORM